Mathematical Analysis for Engineering and Applied Sciences

The book explores a range of mathematical topics essential for application in engineering and applied sciences. It explores both the theoretical and practical aspects, providing a comprehensive foundation for the development of robust theories applicable to engineering and applied sciences.

Mathematical Analysis for Engineering and Applied Sciences: Foundational and Fundamental Aspects discusses the essential mathematical principles that underpin the fields of applied science and engineering. This comprehensive book explores a blend of pure and applied mathematics, demonstrating how mathematical tools and techniques can be utilized to create a wide range of models for practical applications in these disciplines. It addresses the challenges of handling complex phenomena and provides algorithms, methods, and logical concepts that are invaluable for bioengineering, cryptosystems, surface modeling, and various other engineering applications.

Individual researchers, educators, students, and department libraries will find this book of interest.

Mathematics and its Applications
Modelling, Engineering, and Social Sciences
Series Editor: Hemen Dutta, Department of Mathematics, Gauhati University

Sequence Spaces: Topics in Modern Summability Theory
Mohammad Mursaleen and Feyzi Başar

Fractional Calculus in Medical and Health Science
Devendra Kumar and Jagdev Singh

Topics in Contemporary Mathematical Analysis and Applications
Hemen Dutta

Sloshing in Upright Circular Containers: Theory, Analytical Solutions, and Applications
Alexander Timokha and Ihor Raynovskyy

Advanced Numerical Methods for Differential Equations: Applications in Science and Engineering
Edited by Harendra Singh, Jagdev Singh, S. D. Purohit, and Devendra Kumar

Concise Introduction to Logic and Set Theory
Edited by Iqbal H. Jebril, Hemen Dutta, and Ilwoo Cho

Integral Transforms and Engineering: Theory, Methods, and Applications
Abdon Atangana and Ali Akgül

Numerical Methods for Fractal-Fractional Differential Equations and Engineering; Simulations and Modeling
Muhammad Altaf Khan and Abdon Atangana

Engineering Mathematics and Artificial Intelligence: Foundations, Methods, and Applications
Edited by Herb Kunze, Davide La Torre, Adam Riccoboni, Manuel Ruiz Galán

Mathematical Analysis for Engineering and Applied Sciences: Foundational and Fundamental Aspects
Edited by Hemen Dutta and Ahmet Ocak Akdemir

For more information about this series, please visit: www.routledge.com/Mathematics-and-its-Applications/book-series/MES

ISSN (online): 2689–0224
ISSN (print): 2689–0232

Mathematical Analysis for Engineering and Applied Sciences

Foundational and Fundamental Aspects

Edited by Hemen Dutta and
Ahmet Ocak Akdemir

CRC Press is an imprint of the
Taylor & Francis Group, an informa business

Designed cover image: Web Large Image (Public); Getty Images

First edition published 2025
by CRC Press
2385 NW Executive Center Drive, Suite 320, Boca Raton FL 33431

and by CRC Press
4 Park Square, Milton Park, Abingdon, Oxon, OX14 4RN

CRC Press is an imprint of Taylor & Francis Group, LLC

Library of Congress Cataloging-in-Publication Data
Names: Dutta, Hemen, 1981– editor. | Akdemir, Ahmet Ocak, 1985– editor.
Title: Mathematical analysis for engineering and applied sciences : foundational and fundamental aspects / edited by Hemen Dutta and Ahmet Ocak Akdemir.
Description: First edition. | Boca Raton, FL : CRC Press, 2025. | Series: Mathematics and its applications: modelling, engineering, and social sciences | Includes bibliographical references and index.
Identifiers: LCCN 2024032453 (print) | LCCN 2024032454 (ebook) | ISBN 9781032344843 (hardback) | ISBN 9781032345758 (paperback) | ISBN 9781003322856 (ebook)
Subjects: LCSH: Engineering mathematics. | Mathematical models.
Classification: LCC TA331 .M385 2025 (print) | LCC TA331 (ebook) | DDC 620.001/51—dc23/eng/20241122
LC record available at https://lccn.loc.gov/2024032453
LC ebook record available at https://lccn.loc.gov/2024032454

ISBN: 978-1-032-34484-3 (hbk)
ISBN: 978-1-032-34575-8 (pbk)
ISBN: 978-1-003-32285-6 (ebk)

DOI: 10.1201/9781003322856

Typeset in Times
by Apex CoVantage, LLC

Contents

Preface

This book covers a number of potential mathematical topics of a foundational and fundamental nature with significant relevance in applied sciences and engineering. It focuses on both the pure and applied aspects of the topics, as both are important in developing rich theory for prospective use in applied sciences and engineering. The book contains several new algorithms, methods, approaches, and ideas, as well as judicial coverage of relevant theory and applications.

Chapter 1, "Pure Mathematics Applied to Bio-Engineering Problems", describes bio-engineering applications based on pure mathematics. The applications include intuitionistic fuzzy logic and generalized Petri net medical decisions and number theory in modeling infectious diseases, curve recognition in diabetes, and a gear mechanism in an X-ray camera.

Chapter 2, "Inverse Coefficient Problem for Nonlinear Euler-Bernoulli Equation With Periodic Boundary and Integral Addition Conditions", investigates the solution to the inverse problem of a nonlinear Euler-Bernoulli equation with a periodic boundary and integral over determination conditions. Picard's successive approximation theorem and the Fourier method are used to demonstrate the existence, uniqueness, and continuous dependence of the solution on the input data under certain natural regularity and consistency conditions. In addition, a numerical procedure is provided using the linearization method and the finite difference method.

Chapter 3, "A Classification of Focal Surfaces of a Tube Surface in $\mathbb{E}^3$", discusses the focal surfaces of a tube surface in three-dimensional Euclidean space $\mathbb{E}^3$. The tube surface parametrizations are described in relation to Darboux and Frenet frames, yielding focal surfaces for these types of tube surfaces. It draws certain implications about these surfaces becoming flat and how focal surfaces of a tube surface cannot be minimal in $\mathbb{E}^3$. It also evaluates several focal surface instances provided by Frenet and Darboux frames, concluding that s-parameter curves cannot be focal surface asymptotic curves.

Chapter 4, "Approximation of Functions in Certain Lipschitz Classes by (M, λ_n) $(E, 1)$ Means of Fourier Series and Conjugate Series of Fourier Series", defines a product summability method by superimposing the (M, λ_n) method on the $(E, 1)$ method. The method is applied to obtain the degree of approximation of a Fourier series of functions as well as a conjugate series of a Fourier series of functions in certain Lipschitz classes.

Chapter 5, "NTRU Cryptosystem Over Rational Numbers Q_p and a New Verification Method Over Rational Functions Field", discusses the NTRU cryptosystem in terms of rational numbers. With the algebraic contributions of the field structure, the system is built on a broad substructure, allowing for larger message sets and specific key selection.

Chapter 6, "Metric Space and Fixed-Point Results in G-Metric Spaces", discusses the extended properties of G-metric spaces and fixed-point results of generalized contractive mappings in the context of G-metric spaces. Fixed-point results for generalized contractive mappings are also developed on partially ordered G-metric spaces. T-Hardy Rogers–type contractive mappings are defined in the context of ordered G-metric spaces, and their fixed point results are provided. Furthermore, the periodic point property of contractive mappings is determined.

Chapter 7, "Inertial Three-Step Forward-Backward Splitting Algorithms to Solve Inclusion Problems and Their Application to Image Restoration Problems", proposes an advanced inertial three-step forward-backward splitting algorithm based on the inertial technique to solve the inclusion problem of the sum of monotone operators. The proposed algorithm is applied to solve the unconstrained image restoration problems and the compare convergence rate of the studied iteration with the modified viscosity iterative method and Douglas-Rachford and classical forward-backward algorithms.

Chapter 8, "Explicit Study of Fractal Generation of Mandelbrot and Julia Sets via Picard-S Iteration Scheme With s-Convexity", aims to generate the Mandelbrot and Julia sets using a Picard-S iteration scheme with s-convexity. Many graphics of Julia and Mandelbrot sets of the proposed three-step iterative process with s-convexity are presented.

The editors are grateful to the contributors, reviewers, and publishing staff for their cooperation and assistance throughout the process. The editors also express gratitude to their family members, friends, and colleagues for inspiring them to pursue this scholarly work.

Editors
December, 2024

Editors

Hemen Dutta has been a faculty member at Gauhati University, India, since 2010 and presently holds the position of associate professor. Before joining Gauhati University, he served three other academic institutions in different capacities. He is currently interested in applied analysis and mathematical modeling. Dr. Dutta has published over 180 publications as research papers, book chapters and proceedings papers. He has also published several authored and edited books.

Ahmet Ocak Akdemir is a professor in the Department of Mathematics at Ağrı Ibrahim Çeçen University, Ağrı, Turkey. His research interests focus mainly on inequality theory, convex analysis, and real functions of two variables, especially fractional calculus, and integral operators. He has published several research papers with pioneer journals and has delivered several talks at international conferences and meetings. Dr. Akdemir has organized several international conferences as chairman and member of the organizing committee. He is the recipient of numerous publication encouragement awards given by his university as well as private institutions.

Contributors

Mujahid Abbas
Government College University
Lahore, Pakistan
and
University of Pretoria
Pretoria, South Africa

Yeşim Aküzüm
Kafkas University
Kars, Turkey

İrem Bağlan
Kocaeli University
Kocaeli, Turkey

Sezgin Büyükkütük
Kocaeli University
Kocaeli, Turkey

W. Cholamjiak
University of Phayao
Phayao, Thailand

Shilpa Das
Gauhati University
Guwahati, India

S. Das
Kamrup College
Chamata, Nalbari, India

Ömür Deveci
Kafkas University
Kars, Turkey

Hemen Dutta
Gauhati University
Guwahati, India

Fatma Kanca
Fenerbahçe University
İstanbul, Turkey

İlim Kişi
Kocaeli University
Kocaeli, Turkey

Talat Nazir
University of South Africa
Johannesburg, South Africa

Günay Öztürk
Izmir Democracy University
İzmir, Turkey

Shikha Pandey
Jaypee Institute of Information Technology
Noida, U.P., India

Mehmet Sever
Kilis 7 Aralık University
Kilis, Turkey

Nisha Sharma
Pt. J.L.N. Govt. College
Faridabad, India

Anthony G. Shannon
University of New South Wales
Kensington, Australia

D. Yambangwai
University of Phayao
Phayao, Thailand

1 Pure Mathematics Applied to Bio-Engineering Problems

Anthony G. Shannon, Ömür Deveci, and Yeşim Aküzüm

1.1 INTRODUCTION

Many of the great bio-medical discoveries of the last two centuries have been serendipitous [1] and driven by curiosity-driven and basic research [2], both of which, along with expository writing, are unfashionable in an era of rankings of universities in research league tables [3]. In this chapter we consider some examples of the role of pure mathematics in contributing to research in biological and medical applications [4]. The salient features in each example will be explained to convey the essence of the pure mathematics involved. The references in the examples are readily accessible on the relevant websites so that the interested reader can see where the message proceeds from the outline presented here.

The examples chosen in this chapter also illustrate the serendipitous paths that much experimental work actually takes in practice, which can be very different from the neat systematics arrangements in texts on the philosophy of science [5]. The relative emphases of the four otherwise unrelated bio-engineering examples can also be displayed on a two-way grid, as in Figure 1.1.

The initial case considers the role of intuitionistic fuzzy logic (IFL) [6] and generalized nets [7] in decision-making in medicine, a field riddled with uncertainty [8]. There is a sense [9] in which IFL integrates the intuitionism of Brouwer [10] and the fuzziness of Zadeh [11].

The second case involves Fibonacci numbers, which were part of possibly the first mathematical model in biology by Leonard Filius Bonacci himself in the thirteenth century. The relevance to diabetes arises from the apparent role of infectious diseases in the onset of an auto-immune disease [12, 13].

Third, we have an example of scientific exploration as it frequently happens. In this case, it arose from viewing the decay curves of insulin, C-peptide, and glucose after oral glucose tolerance tests on thousands of patients. These curves looked more like the gamma-variate curve than the Gaussian curve, and the evidence of the coefficient of determination when applied to the data seemed to confirm this [14]. The next step was to look at the pure mathematics, followed by a possible medical explanation. This involved the fact that the pro-insulin molecule is cleaved into

DOI: 10.1201/9781003322856-1

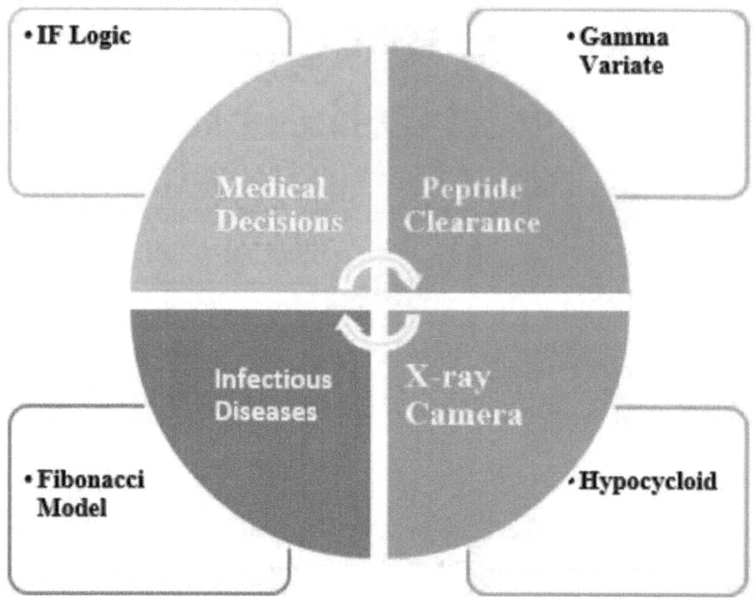

FIGURE 1.1 Management (vertically increasing), modeling (horizontally increasing).

insulin and C-peptide as the stimulated chemicals proceed through the portal vein into the liver, where a generally unknown proportion of the insulin is stored. Then more sophisticated statistical measures needed to be applied so that the guesswork could be taken out of estimating the pre-hepatic secretion of insulin in the person with Type 2 diabetes.

The fourth and final case here involves the application of a hypocycloid gear mechanisms [15]. Hypocycloid and epicycloid examples adorn nearly every first course in the differential calculus in universities but very rarely with examples of any applications. (An epicycloid gear mechanism is outlined in [16].)

1.2 INTUITIONISTIC FUZZY LOGIC IN MEDICAL DECISION MAKING

1.2.1 Introduction

The process of diagnosis is crucial to medical practice. The process of determining the nature of a problem is called diagnosis, and it involves taking into account the patient's signs and symptoms and medical history and, if necessary, the findings of laboratory tests and other diagnostic procedures.

In an environment of uncertainty, the physician takes decisions. He or she decides whether further tests or treatments are necessary and estimates the possibility of a disease. The history, the pertinent physical examination, the findings of laboratory testing, and the outcomes of additional diagnostic hypothesis all consist of information that is utilized to change the diagnostic hypothesis. Thus, on the basis of information gathered at any moment about the patient, a decision on the subsequent movement of the patient is determined.

Not infrequently there is a lack of precise information (e.g., history of birth trauma, kernicterus), or the information is ambiguous (e.g., results in the upper or the lower limits of the normal), uncertain (e.g., family history), or doubtful (e.g., history of rheumatic fever). In this section we focus on some places in a generalized net (GN) for chorea where the decision is arbitrary because between the distinct extreme positions (e.g., "positive–negative"), many intermediate positions with varying degrees of membership are possible.

Chorea is a jerky involuntary movement particularly affecting the head, face, or limbs. Athetosis is an uncontrollable writhing movement that primarily affects the hands, face, and tongue. Both are representatives of a dyskinesia—a group of involuntary movements that appear to be fragmentation of the normal smoothly controlled limb and facial movements. Most patients have both chorea and athetosis. Here we shall illustrate the possibility for implementation of the elements of intuitionistic fuzzy logic in decision making in medicine constructing a GN containing elements of the IFL [8].

1.2.2 Intuitionistic Fuzzy Logic

Each proposition (in the traditional sense) can be given a truth value, either 1 for truth or 0 for falsity. This truth value is a real number in the range [0, 1] and is referred to as the "truth degree" of a certain proposition in fuzzy logic. Here, we add a further value called "falsity degree" that falls inside the range [0, 1]. As a result, the proposition p is given two real numbers, $\mu(p)$ and $v(p)$, with the requirement that it hold:

$$\mu(p)+v(p)\leq 1$$

Each GN condition transition predicate will be approximated in the aforementioned (intuitionistic fuzzy) way. The decision-making process in a neurological disorder associated with chorea and athetoid motions is the focus of this GN. Here, we display the GN positions (see Figure 1.2).

The tokens enter the GN with an initial characteristic "findings of chorea and athetoid movements".

$$Z_1 = < \{l_1\}, \{l_2, l_3\}, \begin{array}{c|cc} & l_2 & l_3 \\ \hline l_1 & W_{1,2} & W_{1,3} \end{array} >$$

in which $W_{1,2}$ = "the onset is in childhood", and $W_{1,3}$ = "the onset is in adulthood". The tokens do not have any characteristic in places l_2 and l_3. In the assessment

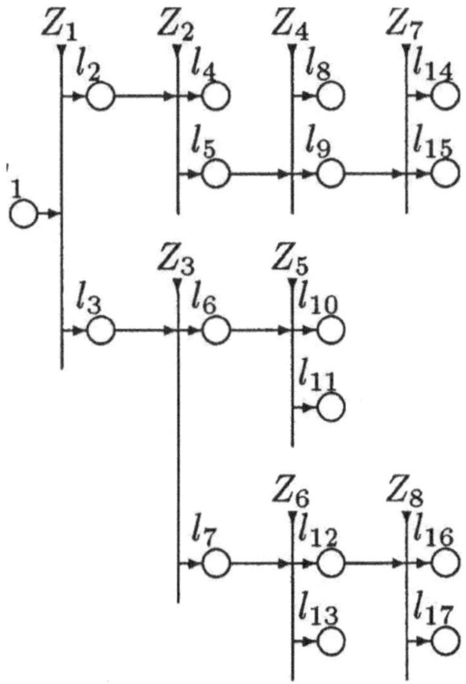

FIGURE 1.2 GN for chorea.

of the onset (childhood or adulthood) there may be some degree of doubt, such as whether a 17-year-old patient must be considered a child or an adult.

$$Z_2 = < \{l_2\}, \{l_4, l_5\}, \begin{array}{c|cc} & l_4 & l_5 \\ \hline l_2 & W_{2,4} & W_{2,5} \end{array} >$$

where $W_{2,4}$ = "there is a history of rheumatic fever", and $W_{2,5} = W_{2,4}$. The tokens then take on the characteristic "the diagnosis is: Sydenham's chorea" in place l_4, and they do not have any characteristic in place l_5. Frequently the course of the rheumatic fever is clinically oligo- or asymptomatic. It should always be dubious if the patient has no real history of rheumatic fever (which can actually be a trigger for some patients in the development of Type 1 diabetes mellitus, discussed later in this chapter).

$$Z_3 = < \{l_3\}, \{l_6, l_7\}, \begin{array}{c|cc} & l_6 & l_7 \\ \hline l_3 & W_{3,6} & W_{3,7} \end{array} >$$

where $W_{3,6}$ = "the family history is positive", and $W_{3,7} = W_{3,6}$.

Now the tokens take on the characteristic "Huntington and/or Wilson's disease(s) are possible" in place l_6, and they do not have any characteristic in place l_7. The patient's lack of knowledge of family history of some disease is a typical example of the uncertainty of the obtained information, that is, of information that is unreliable.

$$Z_4 =< \{l_5\}, \{l_8, l_9\}, \begin{array}{c|cc} & l_8 & l_9 \\ \hline l_5 & W_{5,8} & W_{5,9} \end{array} >$$

Here $W_{5,8}$ = "there is history of viral infection", $W_{5,9} = W_{5,8}$. The tokens take on the characteristic "postencephalitic symptoms are present" in place l_8, but they do not have any characteristic in place l_9.

$$Z_5 =< \{l_6\}, \{l_{10}, l_{11}\}, \begin{array}{c|cc} & l_{10} & l_{11} \\ \hline l_6 & W_{6,10} & W_{6,11} \end{array} >$$

where $W_{6,10}$ = "Kayser-Fleischer ring is present", and $W_{6,11} = W_{6,10}$.

The tokens gain the characteristics "the diagnosis is: Wilson's disease; treat with penicillamine and low-copper diet" in place l_{10} and "the diagnosis is Huntington's disease; genetic researching is necessary; treat chorea and behavioral disturbances (e.g., psychoses, depression)" in place l_{11}.

$$Z_6 = \{l_7\}, \{l_{12}, l_{13}\}, \begin{array}{c|cc} & l_{12} & l_{13} \\ \hline l_7 & W_{7,12} & W_{7,13} \end{array} >$$

in which $W_{7,12}$ = "the occurrence is unilateral", and $W_{7,13}$ = "the occurrence is bilateral".

The tokens do not have any characteristic in place l_{12}, and they gain the characteristic "the possible diagnosis is: encephalitis (CSF findings), or pregnancy, or metabolic disorder (hyperthyroidism, hypoparathyroidism, acanthocytosis), or autoimmune disorder (systemic lupus erythematosus), or drug reaction (L-dopa, phenytoin), lithium, phenothiazines, oral contraceptives, carbamazepine and pemoline" in place l_{13}.

$$Z_7 = \{l_9\}, \{l_{14}, l_{15}\}, \begin{array}{c|cc} & l_{14} & l_{15} \\ \hline l_9 & W_{9,14} & W_{9,15} \end{array} >$$

in which $W_{9,14}$ = "there is a history of birth trauma", and $W_{9,15} = W_{9,14}$.

The tokens have the characteristics “the diagnosis is: kernicterus” in place l_{14} and “consider Wilson’s disease” in place l_{16}. Oftentimes, the history of birth trauma or neonatal jaundice (related to breast milk) complicates the course (cause, degree of expression, and duration) as a possible cause for kernicterus.

$$Z_8 = \langle \{l_{12}\}, \{l_{16}, l_{17}\}, \begin{array}{c|cc} & l_{16} & l_{17} \\ \hline l_{12} & W_{12,16} & W_{12,17} \end{array} \rangle$$

where $W_{12,16}$ = “the onset is sudden”, and $W_{12,17}$ = “the onset is gradual”.

Finally, the tokens can have the characteristics “the possible diagnosis is: vascular disorder (hemorrhage or infarct)”; CT scan is necessary; go to NGN55 and NGN57 in [17]) in place l_{16} and “metastases and focal encephalitis are possible”; CT scan is necessary; go to NGN85 in [17]) in place l_{17}: the details are not pertinent so much as illustrating that the decision processes can continue until a decision is reached with the desired efficacy and efficiency [18].

This outline of an intuitionistic fuzzy application is not intended to teach the approach but rather to sensitize the reader to its variety of possibilities when black-and-white decisions may not be wise [19].

1.3 FIBONACCI MODEL OF INFECTIOUS DISEASE

1.3.1 Introduction

One of the earliest mathematical studies of population increase was the Fibonacci rabbit population model, which is frequently recognized as pioneering. Voltera later presented an analytical model of population dynamics [15]. In revisiting the Fibonacci rabbit growth model, Dubeau [20] created a method that can be used in population dynamics and epidemiology, where censoring happens either by death or incapacitation to reproduce. This section’s goal is to integrate [7] with Dubeau’s technique and apply it to Makhmudov’s [12] Fibonacci model of infectious illnesses.

1.3.2 The Model

Following Makhmudov and in light of the COVID-19 pandemic, three epidemiological stages can be hypothesized in the process of an infectious disease spreading:

1. an initial (incubation) phase of r periods (periods $0, 1, 2, \ldots, r-1$) during which people who have the disease do not spread it to others,
2. a period (periods $r, r+1, \ldots, r+t-1$) that has reached its mature (infectious) stage when each person infects s healthy people, and
3. a removal stage of m periods (periods $r+I, \ldots, r+I+m-1$) during which infected individuals are no longer infectious.

A common cold, for example, takes about two days to develop on average, so $r = 2$, and a person is contagious for about three days ($t = 3$), with symptoms lasting about seven days ($r + t + m = 7$, hence $m = 2$). s is generally a variable, but since we don't have any more information, we'll consider it a constant. These correspond to the following in turn as a variation to Fibonacci's rabbit problem:

1. the stage of infancy,
2. the stage of reproduction, and
3. the stage after reproduction,

respectively. For the original Fibonacci model, $r = 2$, $s = 1$, and $t = m + \infty$.

According to Dubeau, let u_n represent the overall number of people that are ill at the nth period and v_n^i represent the number of people that are ill who are older than the i-period at this n^{th} period. For more certainty, v_n^i represents the disease carriers in the

1. initial stage for $i = 0, ..., r-1$,
2. mature stage for $i = r, ..., r+t-1$,
3. removal reproductive stage for $i = r+t, ..., r+t+m-1$,

and for $i = r+t+m, ...,$ v_n^i represents the disease carriers who have been infected in the past but have already recovered. It is convenient to define u_n and v_n^i for all $n \in Z = (..., -3, -2, -1, 0, 1, 2, 3, ...\}$ and $i \in N = (0, 1, 2, 3, ...\}$. We consider the following initial conditions on v_n^i:

$$v_n^i = \begin{cases} 0, & for \begin{cases} n < 0\, and\, i = 0,1,2,\ldots \\ n = 0\, and\, i = 1,2,3,\ldots \end{cases} \\ 1\left(or\, v_0^0\right) & for\, n = 0\, and\, i = 0 \end{cases}$$

Consequently, for any $n \in Z$,

$$v_n^i = v_{n-1}^{i-1} \quad for\, i > 0$$

and

$$v_n^0 = \begin{cases} 1 & for\, n = 0 \\ s\left\{v_n^r + \cdots + v_n^{r+t-1}\right\} & for\, n \neq 0 \end{cases}$$

We obtain

$$v_n^0 = v_{n-1}^0 + s\left\{v_{n-r}^0 - v_{n-r-t}^0\right\} \tag{1.1}$$

for $n > 1$.

Considering the definition provided, for any $n \in Z$

$$u_n = \sum_{i=0}^{r+t+m-1} v_n^i$$

It follows that

$$u_n = \begin{cases} 0 & for\, n < 0 \\ u0 + s\sum_{k=r}^{r+t-1} u_{n-k} & for\, n = 0,\ldots,r+t+m-1 \\ s\sum_{k=r}^{r+t-1} u_{n-k} & for\, n \geq r+t+m \end{cases} \tag{1.2}$$

or

$$u_n = u_{n-1} - \delta_{n,r+t+m} u_0 + s\{u_{n-r} - u_{n-t}\} \tag{1.3}$$

for $n \geq 1$, where $\delta_{i,j} = 0$ if $i \neq j$, or 1 if $i = j$.

Let $\{u_n\}_{n=0}^{+\infty}$ and $\{\tilde{u}_n\}_{n=0}^{+\infty}$ be the sequences generated with m and $m+1$ for the same values of r, t, and s. From (1.3) we have, for $n \geq 1$,

$$u_n = u_{n-1} - \delta_{n,r+t+m} u_0 + s\{u_{n-r} - u_{n-r-t}\}$$

and

$$\tilde{u}_n = \tilde{u}_{n-1} - \delta_{n,r+t+m} u_0 + s\{\tilde{u}_{n-r} - \tilde{u}_{n-r-t}\}$$

Let $\Delta_n u = u_n - \tilde{u}_n$; then from (1.3)

$$\Delta_n u = \Delta_{n-1} u + (\delta_{n,r+t+m} - \delta_{n,r+t+m+1}) u_0 + s\{\Delta_{n-r} u - \Delta_{n-r-t} u\}$$

It follows from (3.1) that

$$\Delta_{r+t+m+n} u = v_n^0 \text{ for } n \geq 0$$

In accordance with Klarner [4], let's have a look at the sequence of $(r+t+n)$-vectors, $\{v_n\}_{n=0}^{+\infty}$:

$$v_n = \left[v_n^0, v_n^1, \ldots, v_n^{r+t+m-1}\right](n = 0,1,2,\ldots)$$

By the equation, they are related

$$v_{n+t} = vn\,F = \cdots = v0\,F^{n+1} \tag{1.4}$$

where $v0=[1,0,\ldots,0]$, and $F=(f_{ij})$ is a square matrix of order $r+t+m$ with entries $f_{ij}\ (i=0,\ldots,r+t+m-1;\ \ j=0,\ldots,r+t+m-1)$ such that

$$v_{n+1}^{j}=\sum_{i=0}^{r+t+m-1} v_{n}^{j}f_{ij}$$

For our context,

$$f_{i0}=\begin{cases} s & for\, i=r,\ldots r+r-1 \\ 0 & elsewhere \end{cases}$$

and for $j=1,\ldots,r+t+m-1$,

$$f_{ij}=\begin{cases} 1 & for\, i=j-1 \\ 0 & elsewhere \end{cases}$$

The characteristic polynomial of F is

$$det(xI-F)=x^{r+t+m}-s\left(x^{t+m}+\cdots+x^{l+m}\right)=c_F(x)$$

The matrix F satisfies its characteristic equation according to the Cayley-Hamilton theorem, and we have $c_F(F)=0$. So, $F^n c_F(F)=0$ for any $n\geq 0$. Thus, it follows that

$$F^n-s\left(F^{n-r}+\cdots+F^{n-(r+t-1)}\right)=0 \text{ for } n\geq r+t+m$$

Finally, from (1.4), we have

$$v_n-s\left(v_{n-r}+\cdots+v_{n-(r+t-1)}\right)=0$$

and, since $u_n = v_n$ 1, where $1=[1,..,1]^T$,

$$u_n-s\left(u_{n-r}+\cdots+u_{n-(r+t-1)}\right)=0$$

for $n\geq r+t+m$.

1.3.3 Limit of Ratios u_{n+1}/u_n

We now consider the difference equation (1.2) of order $r+t-1$:

$$u_n=s\sum_{k=r}^{r+t-1}u_{n-k}\left(n\geq r+t+m\right)$$

If we suppose that the variables u_{m+1}, u_{m+2}, . . . , $u_{r+t+m-1}$ are known, then the sequence $\{n\}_{n=r+t+m}^{+\infty}$ is fully defined. We see that the finite sequence $\{u_n\}_{n=0}^{r+t+m-1}$

for our model (1.2) or (1.3) is a sequence of nondecreasing integers with $u_0 = 1$ (or any initial value $u_0 > 0$). For the investigation of the ratios u_{n+1}/u_n, we take into account the instances $t = 1$ and $t > 1$.

1.3.4 The Case $t = 1$

We have

$$u_n = su_{n-r}\left(n \geq r+m+1\right)$$

It follows that

$$\frac{u_{n+r+1}}{u_{n+r}} = \frac{u_{n+1}}{u_n}\left(n \geq m+1\right)$$

and the sequence of ratios u_{n+1}/u_n is a sequence that is repeated an infinite number of times with length r.

It is completely characterized by the finite sequence

$$\frac{u_{n+1}}{u_n} \text{for } n = m+1,\ldots,m+r$$

Using (1.2), with $u_0 = 1$ as the starting value, we have

$$u_n = \sum_{i=0}^{k} s^i \text{ for} \begin{cases} n \leq r+m, \; and \\ kr \leq n < \left(k+1\right)r \end{cases}$$

and

$$u_n = \sum_{i=0}^{\frac{n}{r}} s^i \text{ for } n = 0,\ldots,r+m$$

Let

$$\rho_U = \frac{r+m}{r} \text{ and } \rho_L = \frac{1+m}{r}$$

then $\rho_U = \rho_L$ or $\rho_L + 1$. Hence, the sequence $\left\{\frac{u_{n+1}}{u_n}\right\}_{n=m+1}^{m+r}$ is such that

$$\frac{u_{n+1}}{u_n} = \begin{cases} 1 & r-2 \text{ times} \\ s\sum_{i=0}^{\rho_L} s^i / \sum_{i=0}^{\rho_U} s^i & 1 \text{ time} \\ \sum_{i=0}^{\rho_U} s^i / \sum_{i=0}^{\rho_L} s^i & 1 \text{ time} \end{cases}$$

It can be shown that the set

$$\left\{\frac{u_{n+1}}{u_n} \mid n = m+1,\ldots,m+r\right\}$$

converges to the set $\{1, s\}$ when m goes to $+\infty$.

1.3.5 The Case $t > 1$

Suppose $K = r + t - 1$. According to the following linear difference equation of order K, the linear difference equation (1.2) is equivalent to

$$u_{n+K} = s\sum_{k=0}^{t-1} u_{n+k} \qquad (n \geq 0)$$

if the sequence $\{u_n\}_{n=0}^{K-1}$ corresponds to the sequence $\{u_n\}_{n=m+1}^{m+K}$ for (1.2); that is, the limit of u_{n+1}/u_n for both equations is identical. Additionally, let's review certain terms and findings related to linear difference equations of the form

$$u_{n+K} - b_1\, u_{n+K-1} - \cdots - b_{K-1}\, u_{n+1} - b_K\, u_n 0 \quad (n \geq 0) \tag{1.5}$$

1. The polynomial $\varphi(\lambda) = \lambda^K - b_1\lambda^{K-1} - \cdots - b_k$ is referred to as the characteristic polynomial for (1.5).
2. The characteristic equation for (1.5) is the equation $\varphi(\lambda) = 0$.
3. The solutions $\lambda_1,..,\lambda_\ell$ of the characteristic equation are the characteristic roots,

so that the first result is a standard result about the general solution of (1.5).

Theorem 1.1: Suppose (1.5) has characteristic roots $\lambda_1,..,\lambda_k$ with multiplicities $j_1,.., j_k$, respectively. Then (1.5) has n independent solutions $n^j\lambda_\ell^n$, $j = 0,\ldots, j_\ell - 1$; $\ell = 1,\ldots,k$. Moreover, any solution of (1.5) is of the form

$$u_n = \sum_{\ell=1}^{k}\sum_{j=0}^{j_\ell - 1} \beta_{\ell,j} n^j \lambda_\ell^n \qquad (n \geq 0)$$

where the $\beta_{\ell,j}$ are obtained from the values of u_n, for $n = 0,\ldots,K-1$.

Proof: See, for example, Kelley and Peterson [21].

The next two results depend on the form of (1.5).

Theorem 1.2: Assume the b_i are nonnegative in (1.5).

1. If at least one b_i is strictly positive, then (1.5) has a unique simple characteristic root $\sigma > 0$, and all other characteristic roots of (1.5) have moduli not greater than σ.

2. If the indices of the b_i that are strictly positive have the common greatest divisor 1, then (5) has a unique simple characteristic root $\sigma > 0$, and the moduli of all other characteristic roots of (1.5) is strictly less than σ.

Proof: See Ostrowski [22, pp. 91–92].

Theorem 1.3: If in (1.5) the b_i are nonnegative and $\{u_n\}_{n=0}^{+\infty}$ is a sequence satisfying (5) such that $u_0, u_1, \ldots, u_{K-1}$ are strictly positive, then we have $u_n \geq \alpha\sigma^n \ (n \geq 0)$, where $\alpha > 0$ is given by

$$\alpha = min\left\{\frac{u_n}{\sigma^n} \mid n = 0, \ldots, K-1\right\}$$

Proof: See Ostrowski [22, p. 93].

Since

$$b_i = \begin{cases} 0 & for\, i = 1, \ldots, r-1 \\ s & for\, i = r, \ldots, r+t-1 \end{cases}$$

and the fact that $r, \ldots, r+t-1$ has a common greatest divisor of 1 for $t > 1$, it follows from Theorem 1.2 that (1.5) has a single simple characteristic root that is greater than 0 and that the moduli of all other characteristic roots are smaller than. Let the characteristic roots of (1.5) be $\lambda_1, \ldots, \lambda_k$ and σ; then, using Theorem 1.1,

$$u_n = \beta\sigma^n + \sum_{\ell=1}^{k} \sum_{j=0}^{j_\ell - 1} \beta_{\ell,j} n^j \lambda_\ell^n$$

Moreover, since $u_0 \geq 1$ and $\{u_n\}_{n=0}^{K-1}$ is a nondecreasing sequence, we obtain, from Theorem 1.3, $u_n \geq \alpha\sigma^n$ for

$$\alpha = min\left\{\frac{u_n}{\sigma^n} \mid n = 0, \ldots, K-1\right\}$$

It follows that:

$$\alpha \leq \frac{u_n}{\sigma^n} = \beta + \sum_{\ell=1}^{k} \sum_{j=0}^{j_\ell - 1} \beta_{\ell,j} n^j \left(\frac{\lambda_\ell}{\sigma}\right)^n$$

and taking the limit on both sides, we have $\lim_{n \to +\infty} u_n / \sigma^n = \beta \geq \alpha > 0$ as a consequence of the following lemma.

Lemma 1.1: If $|\rho| < 1$, then $\lim_{n \to +\infty} n^\alpha \rho^n = 0$ for any $\alpha = 0, 1, 2, \ldots$.

Finally,

$$\frac{u_{n+1}}{u_n} = \sigma \frac{u_{n+1} / \sigma^{n+1}}{u_n / \sigma^n}$$

and we obtain $\lim_{n \to +\infty} u_{n+1} / u_n = \sigma$, where σ is the unique positive root of

$$\varphi(x) = x^{r+t-1} - s\sum_{i=0}^{t-1} x^i \quad (t > 1)$$

More realistically, in any real population, the epidemiological status of members is as follows: (1) susceptibles, (2) infected, and (3) resistants. Thus, there is not an unlimited supply of susceptibles.

Let

- N be the total population,
- S_n be the number of susceptibles at the nth period,
- U_n be the number of infected and carriers at the nth period, and
- R_n be the number of resistants at the nth period.

Then $N = S_n + U_n + R_n$ and the initial conditions are $S_0 = N - 1$, $U_0 = 1$, and $R_0 = 0$. Using the previous notation, we have

$$U_n = \sum_{i=0}^{r+t+m-1} v_n^i$$

$$R_n = \sum_{i=r+t+m}^{+\infty} v_n^i = \sum_{i=r+t+m}^{+\infty} v_{n-i}^0 = \begin{cases} 0 & if\, n < r+t+m \\ \sum_{i=r+t+m}^{n} v_{n-i}^0 & if\, n \geq r+t+m \end{cases}$$

and

$$S_n = N - U_n - R_n$$

However, the number of susceptibles is limited, so that

$$v_n^0 = min\left\{ S_{n-1}, s\sum_{i=r}^{r+t-1} v_n^i \right\}$$

and

$$\begin{aligned} S_n &= S_{n-1} - v_n^0 \\ U_n &= U_{n-1} + v_n^0 - v_n^{r+t+m} \\ R_n &= R_{n-1} + v_n^{r+t+m} \end{aligned}$$

For R_n, we have

$$R_n = R_{n-(r+t+m)} + \sum_{i=0}^{r+t+m-1} v_{n-i}^{r+t+m}$$

$$= R_{n-(r+t+m)} + \sum_{i=0}^{r+t+m-1} v^i_{n-(r+t+m)}$$

$$= R_{n-(r+t+m)} + U_{n-(r+t+m)} = N - S_{n-(r+t+m)}$$

It then follows that

$$S_n - S_{n-(r+t+m)} + U_n = 0$$

or

$$U_n = S_{n-(r+t+m)} - S_n$$

and, if $S_n = 0$, then

$$U_n = S_{n-(r+t+m)}$$

Example: Table 1.1 illustrates this model for $r = 2$, $t = 3$, $m = 2$, $s = 1$, and $N = 200$.

TABLE 1.1
Realistic Model

	v_n^i									
$n\downarrow, i\rightarrow$	**0**	**1**	**2**	**3**	**4**	**5**	**6**	U_n	R_n	S_n
0	1	0	0	0	0	0	0	1	0	199
1	0	1	0	0	0	0	0	1	0	199
2	1	0	1	0	0	0	0	2	0	198
3	1	1	0	1	0	0	0	3	0	197
4	2	1	1	0	1	0	0	5	0	195
5	2	2	1	1	0	1	0	7	0	193
6	4	2	2	1	1	0	1	11	0	189
7	5	4	2	2	1	1	0	15	1	184
8	8	5	4	2	2	1	1	23	1	176
9	11	8	5	4	2	2	1	33	2	165
10	17	11	8	5	4	2	2	49	3	148
11	24	17	11	8	5	4	2	71	5	124
12	36	24	17	11	8	5	4	105	7	88
13	52	36	24	17	11	8	5	153	11	36
14	36	52	36	24	17	11	8	184	16	0
15	0	36	52	36	24	17	11	176	24	0
16	0	0	36	52	36	24	17	165	35	0
17	0	0	0	36	52	36	24	148	52	0
18	0	0	0	0	36	52	36	124	76	0
19	0	0	0	0	0	36	52	88	112	0
20	0	0	0	0	0	0	36	36	164	0
21	0	0	0	0	0	0	0	0	200	0

1.4 GAMMA-VARIATE CURVES IN PEPTIDE CLEARANCE

1.4.1 Introduction

Glucose and insulin are the two essential substances in the body's ongoing effort to manufacture energy. Insulin helps glucose enter cells so that it can be converted into energy. Diabetes mellitus can develop when the body's capacity to receive the energy it requires to function correctly is compromised.

1.4.2 Kinetics

In essence, the two main types of diabetes are distinct illnesses with similar symptoms. In both cases, diabetes mellitus is a chronic condition characterized by an excessive concentration of glucose in the blood. Insulin, a hormone produced and secreted by the beta cells of the pancreatic islets of Langerhans, is the main regulator of blood glucose levels. Low insulin levels and/or an abundance of substances that interfere with insulin's ability to work can both result in high blood sugar levels [23].

The disease condition in which the endogenous generation of insulin in the pancreas is abolished is known as type 1 diabetes mellitus (T1DM or IDDM). Insulin from an external source needs to be provided, usually by subcutaneous injection, although there is still insulin release from the pancreas.[1]

Although occasionally insulin treatment is necessary, this form of the condition is typically treated with a combination of diet, exercise, and oral agents.

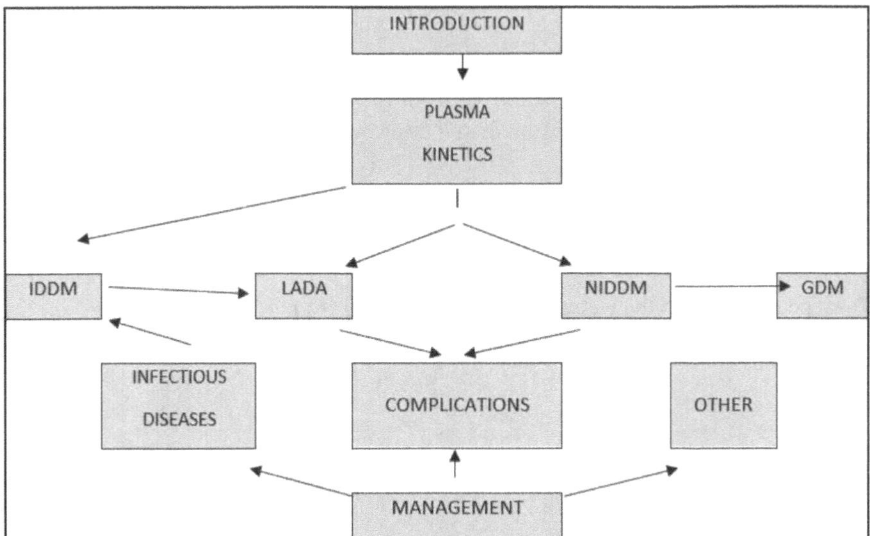

FIGURE 1.3 Diabetes mellitus factors.

Abnormalities in the metabolism of proteins, carbohydrates, and lipids may result from this imbalance. The main complications of diabetes are characteristic symptoms, the progressive development of the kidney and retinal capillary disease, harm to peripheral nerves, and hastened arteriosclerosis.

The main clinical aspects can be emphasized with the use of mathematical modeling of subcutaneous insulin clearance and prehepatic insulin secretion. The process is influenced by variables such as insulin concentration, the half-life of the insulin, injection site, method, and type. The research on pre-diabetes and diabetic complications, as well as the development of therapeutic advances such as pumps, jet injectors, and bio-synthetic human insulins, all benefit from an understanding of insulin kinetics.

We may estimate the pre-hepatic insulin secretion rate in T2DM with compartment [24] and gamma-variate modeling by using the experimental facts that insulin and C-peptide are secreted from the pancreas in equimolar levels and that C-peptide is not deposited in the liver [24].

An initial rising curve shows when observing the presence of insulin in plasma. Since it assumes that all subcutaneous insulin is immediately accessible for transcapillary absorption, this predicts that insulin is supplied to the plasma pool proportionally with time t, though this is an oversimplification of the biological process:

$$y \propto t^a \ (a > 0)$$

or

$$\begin{aligned} \frac{dy}{dt} &= \frac{ay}{t} \quad (t \neq 0) \\ &= k_{sp} x \end{aligned}$$

in which we assume a fractional rate of systemic delivery k_{sp} here and subsequently a degradation rate constant k_d from the subcutaneous compartment. The plasma pool represents the plasma distribution (with k_c as the metabolic clearance rate):

$$\frac{dy}{dt} = \frac{ay}{t} - k_c y$$

Disappearance from the subcutaneous site is found from

$$\begin{aligned} \frac{dx}{dt} &= -\left(k_d + k_{sp}\right) x \\ &= -kx \end{aligned}$$

or

$$x = x_0 e^{-kt}$$

Appearance in the plasma (or clearance from the injection site) is next found from

$$\frac{1}{y}\frac{dy}{dt} - \frac{a}{t} = -k_c$$

$$\frac{d}{dt}\ln y - a\frac{d}{dt}\ln t = -k_c$$

$$\frac{d}{dt}\ln\left(\frac{y}{t^a}\right) = -k_c$$

$$y = y_0 t^a e^{-k_c t}$$

This is a gamma-variate function: from it the prehepatic C-peptide rate can be expressed as:

$$\frac{dy}{dt} = a y_0 t^{a-1} e^{-k_c t} - k_c y_0 t^a e^{-k_c t}$$

where the non-degraded clearance rate from the subcutaneous site is represented by the first term on the right-hand side, and the clearance rate from the plasma pool is represented by the second term. In order to assess insulin absorption, this "gamma-variate" model was tested on patients using the external disappearance of I125-labeled human soluble insulin (U100) and simultaneous measurements of plasma immunoreactive insulin, C-peptide, and glucose. To investigate the relationship between insulin absorption and that metric, 99M Technetium was measured to monitor subcutaneous blood flow [14].

1.4.3 Acknowledgments

The clinical data for this chapter were obtained while working with research grants from

- the Australian Research Council, for work at Prince of Wales Hospital, Sydney,
- the Australian Academy of Sciences, for research at University Hospital Cardiff,
- the Royal Society, for work at the University of Wales College of Medicine,
- the World Health Organization, for a course at Fitzwilliam College, Cambridge,
- the Foundation for Australian Resources, University of Technology Sydney.

The studies in this section were also approved by the local human research ethics committees in each case, and informed consent was given by each subject. Gratitude is also expressed to the following endocrinologists for supervision, discussions, and clinical data provided to the author at various times:

- the late Professor Robert Turner, CBE, University of Oxford, Oxford, England;
- Professor David Owens, CBE, University of Cardiff, Wales;
- Professor Stephen Colagiuri, OA, Faculty of Medicine, University of Sydney, Australia;
- Professor Barry S Thornton, AM, Faculty of Science, University of Technology Sydney.

1.5 HYPOCYCLOID GEAR MECHANISM IN AN X-RAY CAMERA

1.5.1 Introduction

The purpose of this section outlines the mathematical design of the Manners X-ray camera, which employs a hypocycloidal gear system to prevent plane coincidences in a particular form of X-ray crystallography.

In X-ray crystallography, the powder diffraction technique plays a key role. In order to fulfill the Bragg condition for constructive interference of the incident radiation, it is crucial to avoid plane coincidences in practice. There are several approaches that have been devised to achieve this; however, the approach presented here avoids the seemingly straightforward but challenging operation of accurately repositioning the sample in the holder of the Gandolfi camera [11].

It was too complicated to put into practice the obvious theoretical solution of using separate electric motors to control oscillation and rotation. The goal was to create the two movements using a single driving mechanism, avoiding realigning the same crystal surfaces in the process.

1.5.2 The Geometry of the Hypocycloid

The hypocycloid, from Girolamo Cardano (1501–1566), is the location of a fixed point rolling inside the perimeter of a bigger circle without sliding. Any point on the perimeter of the latter circle follows a straight line that passes through the center of the former when the radius of the bigger circle is two times that of the smaller, inner circle; in other words, the diameter. This offered a hint for a suitable mechanism.

Let $P(x, y)$, which was initially located at A, represent the location of the tracing point after the circles' points of contact shifted from A to H. Let θ and φ be the angles through which OH and CH, respectively, have turned, and let R and r be the radii of the fixed and rolling circles, respectively (Figure 1.4).

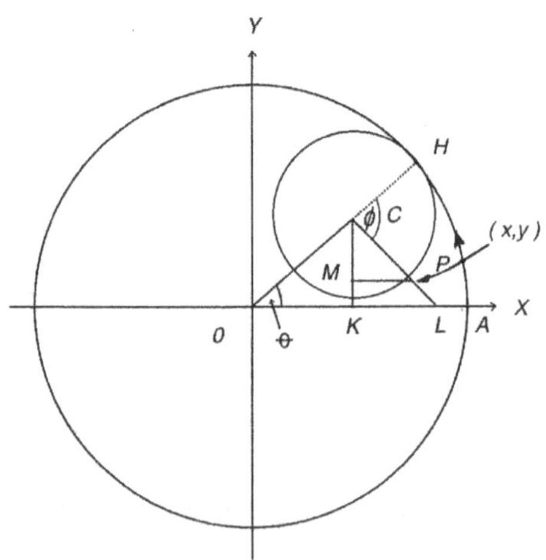

FIGURE 1.4 Hypocycloid.

From Figure 1.5 we can determine the coordinates of point P as follows:

$$\begin{aligned} x &= OK + MP \\ &= (R-r)\cos\theta + r\cos\left(\frac{R-r}{r}\right)\theta \end{aligned}$$

and

$$\begin{aligned} y &= KC - MC \\ &= (R-r)\sin\theta - r\sin\left(\frac{R-r}{r}\right)\theta \end{aligned}$$

which decreases to $y = 0$ when $R = 2r$, and thus P traces out a straight line. So that the bottom of the sample holder may be fixed to this axis and the top fixed to the sample with an axle in between, P can now travel up and down the x-axis.

Wheel 4 in Figure 1.5 and 3 rotates Wheel 1, while Rod 2 goes up and down Slot 5 thanks to a hypocycloidal gear system. In this method, if the rod is attached to the sample support, the latter can rock back and forth as the sample holder is rotated by Wheel 1.

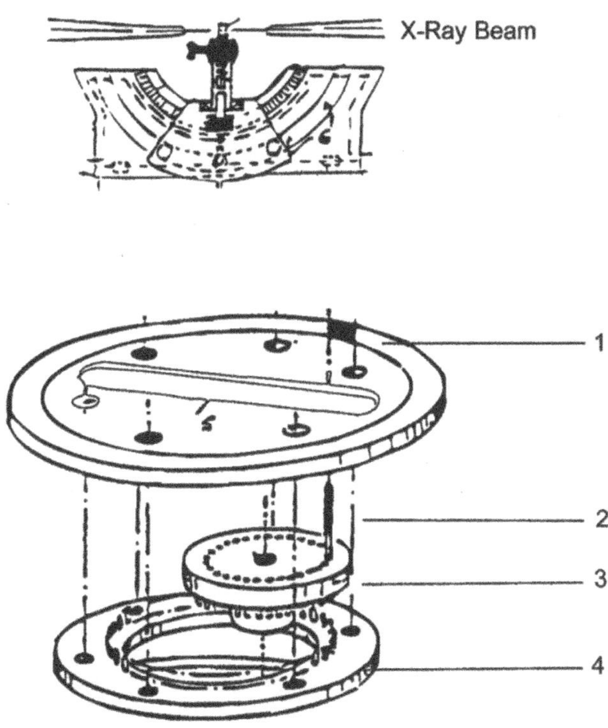

FIGURE 1.5 Hypocycloid gear mechanism.

1.5.3 The Physics behind the Problem

The primary characteristic of the powder diffraction method is the impact of a narrow beam of monochromatic X-rays on crystalline powder made up of fine, randomly oriented particles (as small as 0.1 μm), as shown in Figure 1.6.

For the X-ray beam to consistently identify crystallites with the correct orientation to satisfy the Bragg condition for reflection, all possible orientations of all possible planes must exist.

The Bragg condition for constructive interference of the incident radiation is given by $\sin\psi = n\lambda / 2d$, where d is the distance between successive atomic planes in the crystal, λ is the wavelength of the radiation, and n is an integer (n = 1 for a first-order reflection). The variation in the angle ψ brings different planes into the position of the objectives in the design of the Manners camera, which aims to avoid coincidences of these planes (Figure 1.7).

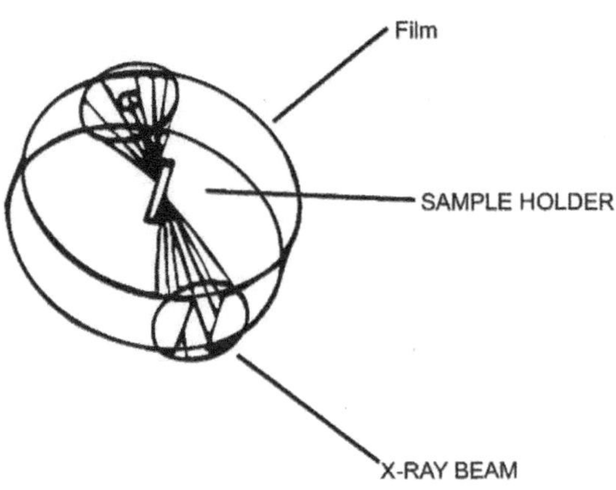

FIGURE 1.6 Gandolfi camera.

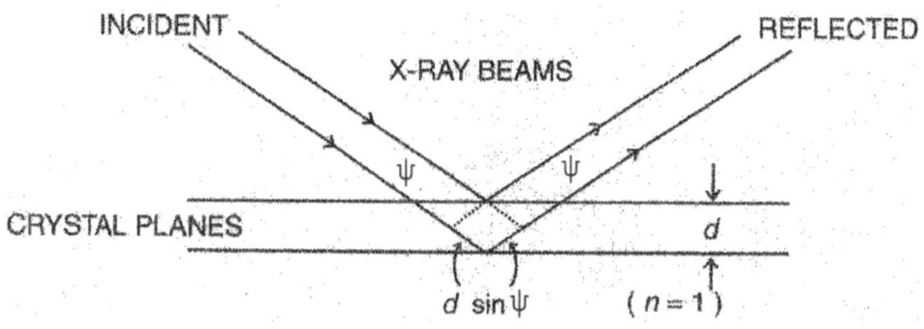

FIGURE 1.7 The Bragg condition.

1.5.4 The Mechanism of the Camera

As Wheel 2 rotates through angle θ, the base of the oscillating mechanism moves along the spherical surface from $P^{/}$ to $P_{\theta}^{/}$. That is (Figure 1.8):

- $P^{/}$: extreme position of oscillating mechanism;
- $P_{\theta}^{/}$: arbitrary position of oscillating mechanism;
- $2\pi/\dot{\phi}$: revolution time of end of rod;
- $2\pi/\dot{\theta}$: time taken for rod to traverse groove;
- *S*: sample (usually a small cylindrical holder filled with powder) is the center of rotation of the rocking mechanism: $(-45^{\circ} < \alpha < 45^{\circ})$ (Figure 1.8).

Thus at P, $\tan 45^{\circ} = R / SP$, so that $SP = \sqrt{2}R$. Then P_{θ} lies on a sphere with center S and radius $\sqrt{2}R$, with the equation

$$x^2 + y^2 + z^2 = 2R^2$$

In Figure 1.4, Wheel 1 rotates with an angular velocity $\dot{\phi}$ relative to small Wheel 2, which rolls inside Wheel 1 (rigidly attached to the base wheel) with an

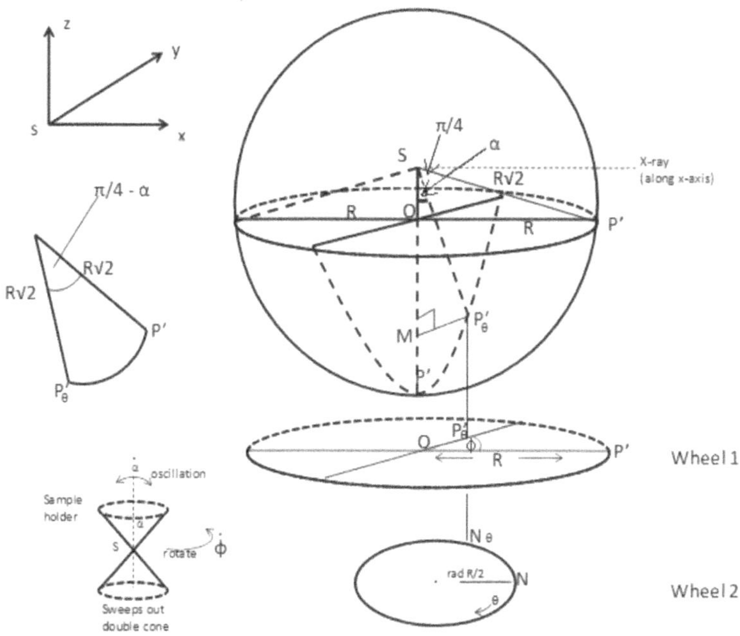

FIGURE 1.8 Geometry of the mechanism.

angular velocity of $\dot{\theta}$ relative to Wheel 2. This causes the rod to move to and fro along the groove, so that if $P_\theta^{/}$ is the position of the rod after Wheel 2 has rotated through an angle θ (relative to Wheel 1), then $P_\theta^{/}(x, y, z)$, a point on the surface of the sphere carved out by the rod top when small Wheel 2 rotates through angle θ relative to Wheel 1, is given by

$$x = R\cos\theta\cos\phi$$
$$y = R\cos\theta\sin\phi$$

We obtain from the sphere's equation

$$\begin{aligned} 2R^2 &= R^2\left(x^2 + y^2 + z^2\right) \\ &= R^2\left(\cos^2\theta\cos^2\phi + \cos^2\theta\sin^2\phi + z^2\right) \\ &= R^2\left(\cos^2\theta\left(\cos^2\varphi + \sin^2\varphi\right) + z^2\right) \\ &= R^2\left(\cos^2\theta + \left(2 - \cos^2\theta\right)\right) \end{aligned}$$

so that

$$z = R\sqrt{2 - \cos^2\theta}$$

In other words, the z-value is unrelated to the necessary rotation of Wheel 1.

The area A swept out is equal to the length of the arc times the thickness of the oscillating mechanism if we assume that h is the thickness of the mechanism, which is equal to the diameter of the cylindrical sample container; that is,

$$A = hR\sqrt{2}\left(\frac{\pi}{4} - \alpha\right)$$

and thus

$$\frac{dA}{dt} = -hR\sqrt{2}\,\frac{d\alpha}{dt}$$
$$\left|\frac{dA}{dt}\right| = hR\sqrt{2}\left|\frac{d\alpha}{dt}\right|$$

In triangle OMP,

$$\begin{aligned} \sin\alpha &= \frac{MP}{SP} \\ &= \frac{z}{R\sqrt{2}} \end{aligned}$$

$$\begin{aligned} z &= R\sqrt{2}\sin\alpha \\ &= R\sqrt{2-\cos^2\theta} \\ 2\sin^2\alpha &= 2-\cos^2\theta \\ \cos^2\theta &= 2-2\sin^2\alpha \\ &= 2\cos^2\alpha \\ \cos\theta &= \sqrt{2}\cos\alpha \end{aligned}$$

While an end point of Groove 5 of length $2R$ does one revolution in $2\pi / \dot{\phi}$ seconds, Rod 2 takes $2\pi / \dot{\theta}$ seconds to traverse the same groove. Coincidences will occur only when the rod is aligned above a position on the wheel that it has occupied previously and also when the wheel is simultaneously aligned with its starting point relative to the X-ray beam. This will occur only when integer multiples of the two angular velocities are equal: $m\dot{\theta} = n\dot{\phi}$. By choosing $(m,n) = 1$, with m a sufficiently large prime, we can, in principle, achieve our original purpose.

To further obviate any risks of alignment, a piezoelectric crystal was employed to provide electronic frequency generation in the direction of the z-axis and so give the holder movement in third dimensions.

1.6 CONCLUDING COMMENTS

The foregoing are isolated examples only. The point is that a mathematical conceptual framework is essential for an engineer to view non-routine problems. Yet there are apparent deficiencies in the so-called STEM subjects in many parts of the world: "Australia's university students also are reluctant to pursue subjects such as physics, chemistry, computer science and engineering at the highest level" [25]. It would go beyond the limitations of this chapter to explain how the difference calculus from discrete mathematics can sometimes be more useful in practice than the differential calculus from continuous mathematics; for instance, when plasma glucose measurements are made at 10-minute intervals, then assuming continuity and linearity can overlook problems of ill conditioning during those intervals!

NOTE

1 There are also latent autoimmune diabetes in adults (LADA) and gestational diabetes mellitus (GDM), which are beyond the scope of what we are attempting.

REFERENCES

[1] Chew, Mabel and Karl Sharrock. 2007. *Medical Milestones: Celebrating Key Advances Since 1840.* London: BMJ Publishing.

[2] Ivison, Duncan. 2021. "For Unis, It Is Basic Research That Allows for Discovery." *The Australian*, 3 March, p. 11.

[3] Larcombe, Peter J. 2019. "On the Notion of Mathematical Genius: Rhetoric and Reality." *Palestine Journal of Mathematics*. 8 (1): 121–126.

[4] Klarner, D.A. 1976. "A Model for Population Growth." *77ie Fibonacci Quarterly*. 14 (3): 277–281.

[5] Sekhon, J.G., A.G. Shannon, C. Chiarella and A.F. Horadam. 1984. "Mathematical Modelling and the Preparation of Industrial Mathematicians." In J.S. Berry, D.N. Burghes, I.D. Huntley, D.J.G. James and A.O. Moscardini (eds.), *Teaching and Applying Mathematical Modelling*. Chichester: Ellis Horwood, pp. 235–245.

[6] Atanassov, Krassimir T. 2017. *Intuitionistic Fuzzy Logics*. Cham, Switzerland: Springer.

[7] Atanassov, Krassimir T. 1991. *Generalized Nets*. Singapore: World Scientific.

[8] Shannon, A.G., S.-K. Kim, Y.H. Kim, J.G. Sorsich, K.T. Atanassov and P. Georgiev. 1997. "A Possibility for Implementation of Elements of the Intuitionistic Fuzzy Logic in Decision Making in Medicine." *Notes on Intuitionistic Fuzzy Sets*. 3 (4): 40–43.

[9] Atanassov, K.T. and A.G. Shannon. 1998. "A Note on Intuitionistic Fuzzy Logics." *Acta Philosophica*. 7 (1): 121–125.

[10] Brouwer, L.E.J. (translated by Arnold Dresden). 1913. "Intuitionism and Formalism." *Bulletin of the American Mathematical Society*. 20 (2): 81–96.

[11] Shannon, A.G. and C.K. Wong. 2010. "Bioprocess Engineering and the Manners-Gandolfi X-ray Camera." *International Journal Bioautomation*. 14 (3): 171–178.

[12] Makhmudov, A. 1983. "On Fibonacci's Model of Infectious Disease." In G.I. Marchuk and L.N. Belyk (eds.), *Mathematical Modelling in Immunology and Medicine*. Amsterdam: North Holland, pp. 319–323.

[13] Shannon, A.G., J.H. Clarke and L.J. Hills. 1988. "Contingency Relations for Infectious Diseases." In M. Witten (ed.), *Mathematical Models in Medicine*, Volume 2. Oxford: Pergamon, pp. 829–833.

[14] Shannon, A.G. 2016. "The Development of Legal Probability: Shades of Guilt." *Metascience* (Springer). 25 (1): 115–118.

[15] Deakin, M.A.B. and D.L.S. McElwain. 1994. "Approximate Solution for an Integro-Differential Equation." *International Journal of Mathematical Education in Science and Technology*. 25 (1): 1–4.

[16] Flores, Mercedes, Maria Victoria Carbonell Padrino and Elvira Martinez. 2011. "Design of Cycloids, Hypocycloids and Epicycloids Curves with Dynamic Geometry Software. Engineering Applications." *3rd International Conference on Education and New Learning Technologies*. Barcelona, Spain, ISSN: 3240–1117.

[17] Atanassov, K., M. Daskalov, P. Georgiev, S. Kim, Y. Kim, N. Nikolov, A. Shannon and J. Sorsich. 1997. *Generalized Nets in Neurology*. Sofia: Marin Drinov Publishing House of the Bulgarian Academy of Sciences.

[18] Shannon, Anthony, Olympia Roeva, Tania Pencheva and Krassimir Atanassov. 2004. *Generalized Nets Modelling in Biotechnological Processes*. Sofia: Marin Drinov Publishing House of the Bulgarian Academy of Sciences.

[19] Dubeau, François and A.G. Shannon. 1996. "A Fibonacci Model of Infectious Disease." *The Fibonacci Quarterly*. 34 (3): 257–270.

[20] Dubeau, F. 1993. "The Rabbit Problem Revisited." *The Fibonacci Quarterly*. 31 (3): 268–274.

[21] Kelley, W.G. and A.C. Peterson. 1991. *Difference Equations: An Introduction with Applications*. San Diego, CA: Academic Press.

[22] Ostrowski, A.M. 1973. *Solution of Equations in Euclidean and Banach Spaces.* New York: Academic Press.
[23] Shannon, A.G. 2005. "Using Glucose Tolerance Tests to Model Insulin Secretion and Clearance." *International Journal Bioautomation.* 2 (1): 93–106.
[24] Chiarella, C. and A.G. Shannon. 1987. "An Example of Diabetes Compartment Modelling." In M. Witten (ed.), *Mathematical Models in Medicine: Diseases and Epidemics*, Volume 1. New York: Pergamon Press, pp. 1239–1244.
[25] Barlow, Thomas. 2020. "Maths Teaching at Schools and Universities Must Lift." *The Australian*, 17 November, p. 20.

2 Inverse Coefficient Problem for Nonlinear Euler-Bernoulli Equation with Periodic Boundary and Integral Addition Conditions

İrem Bağlan and Fatma Kanca

2.1 INTRODUCTION

The Euler-Bernoulli ray problem was studied by D. Bernoulli and L. Euler. $w(x,\eta)$ is the transverse displacement at time t and position x from one end of the beam taken as the origin, $n(x)$ is the flexural rigidity, and $m(x) > 0$ is the lineal mass. The fourth-order Euler differential equation is represented in its most general form as follows:

$$m(x)(\partial^2 w)/(\partial\eta^2) + n(x)(\partial^2 w)/(\partial x^4) = 0, \eta > 0, 0 < x < L$$

The Euler-Bernoulli beam equation (1) is frequently encountered. This equation has been studied in many different ways. For example, Gottlieb studied the inhomogeneous variants of this equation for $m(x) = 1$ and $n(x) = 1$ [1]. Soh examined another approach using symmetry for the same problem for the Lie Euler-Bernoulli ray [2]. Following this, they applied the equivalence method of Morozov and Soh Cartan for this problem. Recently, Ndogmo got the exact equivalence transformations for this problem [3, 4]. Özkaya and Pakdemirli handled the transverse vibrations of a rapidly moving beam with time using the symmetry method and got approximate solutions for a decreasing exponential and harmonic problem [5].

The movement of a loaded thin elastic beam is governed by a dynamic beam below a fourth-order Partial Differential Equation (PDE) is given by:

$$(EIw_{xx})_{xx} + \mu w_{tt} = g(x,\eta), \eta > 0, 0 < x < L \tag{2.1}$$

where E is the elastic modulus, I is the area of inertia, μ is the mass per unit length (E, I, μ as constants), and $w(x,\eta)$ is the transverse displacement at time t and position

 DOI: 10.1201/9781003322856-2

x, f dependent on w. Bokhari dealt with the Euler equation given subsequently with symmetry [6].

$$w_{xxxx} + w_{\eta\eta} = g(x,\eta), \eta > 0, 0 < x < L \tag{2.2}$$

The Euler-Bernoulli equation is used in many common areas. Vibration, buckling, elasticity, nanotechnology (nanotubes, nanofillers for nanomotors, nanobearings, and nanosprings) are modelled by the Euler-Bernoulli equations. Euler's equation can be encountered in other models, for example, the power-law applied load case and the general power law [7–12].

Inverse problems are used to find an unknown property of an object or a place. Inverse problems are especially important for many calculations used in aircraft, missiles, and submarines. In geophysics, the inverse problem is finding subsurface inhomogeneities. In technology, when measuring the frequencies of a piece of material, the inverse problem is finding whether there is a defect (a hole in a metal) in that material. In medicine, there may be a tumor or abnormalities in the human body. In this, the inverse problem is used. It is used in many fields such as population, electrochemistry, engineering, chemistry, and plasma physics. The problems with nonlocal boundary conditions discussed in this chapter are not easy to study. Various boundary conditions have been studied in this area. [13–15, 16–18].

Fourth-order nonlinear equations contribute to the solution of many engineering problems, and many linear and non-linear studies can be found in this field [8–13].

At the same time, periodic boundary conditions are also difficult-to-apply conditions that are frequently used in many areas, for example, heat transfer and life sciences [14, 15, 19].

In this chapter, the Euler-Bernoulli quasilinear problem with periodic boundary conditions is studied.

The chapter consists of four parts:

In 2.2, obtaining the existence and uniqueness of the solution of the following problem is shown by Fourier and Picard's method. In 2.3, the stability of the solution on the data is examined. In 2.4, the numerical approximation is presented.

Let's examine the following problem, the unknown function $\{w(x,\eta), q(\eta)\}$:

$$\frac{\partial^2 w}{\partial \eta^2} + \frac{\partial^4 w}{\partial x^4} = q(\eta) h(x,\eta,w), (x,\eta) \in \Omega \tag{2.3}$$

$$\begin{aligned} w(0,\eta) &= w(\pi,\eta) \\ w_x(0,\eta) &= w_x(\pi,\eta) \\ w_{xx}(0,\eta) &= w_{xx}(\pi,\eta) \\ w_{xxx}(0,\eta) &= w_{xxx}(\pi,\eta), \eta \in [0,T] \end{aligned} \tag{2.4}$$

$$w(x,0) = \vartheta(x), w_\eta(x,0) = \omega(x), x \in [0,\pi] \tag{2.5}$$

$$S(\eta) = \int_0^\pi \alpha w(\alpha,\eta) d\alpha, \eta \in [0,T] \tag{2.6}$$

where $\vartheta(x)$, $\omega(x)$, $S(\eta)$ are given functions, $h(x,\eta,w)$ is the quasilinear term, and $\Omega := \{0 < x < \pi, 0 < \eta < T\}$.

Nomenclature

$\vartheta(x), \omega(x)$ initial condition
$q(\eta)$ inverse unknown coefficient
$S(\eta)$ overdetermination data
$w(x,\eta)$ solution of problem
$h(x,\eta,w)$ function
$w_0(\eta), w_{ck}(\eta), w_{sk}(\eta)$ Fourier coefficients
M_1, M_2, M_3 constants
Ω domain of x, η

2.2 ANALYSIS OF THE PROBLEM

Suppose the following is provided:

(L1) $S(\eta) \in C^2[0,T]$.

(L2) $\vartheta(x) \in C^3[0,\pi], \omega(x) \in C^1[0,\pi], S(0) = \int_0^\pi \alpha\vartheta(\alpha)d\alpha$,

(L3) $h(x,\eta,w)$ satisfies the following conditions in $\bar{\Omega} \times (-\infty, \infty)$

(1) $\left| \frac{\partial^{(n)} h(x,\eta,w)}{\partial x^n} - \frac{\partial^{(n)} h(x,\eta,\tilde{\omega})}{\partial x^n} \right| \le b(x,\eta)|w - \tilde{\omega}|, n = 0,1,2,$

where $b(x,\eta) \in L_2(\Omega), b(x,\eta) \ge 0$,

(2) $h(x,\eta,w) \in C[0,\pi],\ \eta \in [0,T], |h| \le M$

(3) $\int_0^\pi \alpha h(\alpha,\eta,w)d\alpha \ne 0,\ \forall \eta \in [0,T]$.

If we apply the Fourier method to equation (2.1)

$$w_0 = \vartheta_0 + \omega_0\eta + \frac{2}{\pi}\int_0^\eta\int_0^\pi (\eta - \kappa)h(\alpha,\kappa,w)q(\kappa)d\alpha d\kappa$$

$$w_{ck} = \vartheta_{ck}\cos(2k)^2\eta + \frac{\omega_{ck}}{(2k)^2}\sin(2k)^2\eta + \frac{2}{4\pi k^2}\int_0^\eta\int_0^\pi h(\alpha,\kappa,w)q(\kappa)\Lambda_\eta \cos 2k\alpha d\alpha d\kappa$$

$$w_{sk} = \vartheta_{sk}\cos(2k)^2\eta + \frac{\omega_{sk}}{(2k)^2}\sin(2k)^2\eta + \frac{2}{4\pi k^2}\int_0^{\eta}\int_0^{\pi} h(\alpha,\kappa,w)q(\kappa)\Lambda_\eta \sin 2k\alpha d\alpha d\kappa$$

$$\begin{aligned} w(x,\eta) &= \frac{1}{2}\left[\vartheta_0 + \omega_0\eta + \frac{2}{\pi}\int_0^{\eta}\int_0^{\pi}(\eta-\kappa)h(\alpha,\kappa,w)q(\kappa)d\alpha d\kappa\right] \\ &+ \sum_{k=1}^{\infty}\cos 2kx\left[\vartheta_{ck}\cos(2k)^2\eta + \frac{\omega_{ck}}{(2k)^2}\sin(2k)^2\eta\right] \\ &+ \sum_{k=1}^{\infty}\cos 2kx\left[\frac{2}{4\pi k^2}\int_0^{\eta}\int_0^{\pi}h(\alpha,\kappa,w)q(\kappa)\Lambda_\eta\cos 2k\alpha d\alpha d\kappa\right] \\ &+ \sum_{k=1}^{\infty}\sin 2kx\left[\vartheta_{sk}\cos(2k)^2\eta + \frac{\omega_{sk}}{(2k)^2}\sin(2k)^2\eta\right] \\ &+ \sum_{k=1}^{\infty}\sin 2kx\left[\frac{2}{4\pi k^2}\int_0^{\eta}\int_0^{\pi}h(\alpha,\kappa,w)q(\kappa)\Lambda_\eta\sin 2k\alpha d\alpha d\kappa\right] \end{aligned} \tag{2.7}$$

where $\Lambda_\eta = \sin(2k)^2(\eta-\kappa), \vartheta_0 = \frac{2}{\pi}\int_0^{\pi}\varphi(\alpha)d\alpha, \vartheta_{ck} = \frac{2}{\pi}\int_0^{\pi}\varphi(\alpha)\cos 2k\alpha d\alpha,$
$\vartheta_{sk} = \frac{2}{\pi}\int_0^{\pi}\varphi(\alpha)\sin 2k\alpha d\alpha,\ w_0 = \frac{2}{\pi}\int_0^{\pi}\omega(\alpha)d\alpha, \omega_{ck} = \frac{2}{\pi}\int_0^{\pi}w(\alpha)\cos 2k\alpha d\alpha,$
$\omega_{sk} = \frac{2}{\pi}\int_0^{\pi}\omega(\alpha)\sin 2k\alpha d\alpha,\ h_0(\eta,w) = \frac{2}{\pi}\int_0^{\pi}h(\alpha,\eta,w)d\alpha,\ h_{ck}(\eta,w) = \frac{2}{\pi}\int_0^{\pi}h(\alpha,\eta,w)$
$\cos 2k\alpha d\alpha,\ h_{sk}(\eta,w) = \frac{2}{\pi}\int_0^{\pi}h(\alpha,\eta,w)\sin 2k\alpha d\alpha,\ k = 1,2,3,\ldots$

From (L1)–(L3), we get

$$S''(\eta) = \int_0^{\pi}\alpha w_{\eta\eta}(\alpha,\eta)d\alpha, 0 \le \eta \le T \tag{2.8}$$

From (2.8) and (2.9)

$$q = \frac{S''(\eta) - \pi\sum_{k=1}^{\infty}(2k)^3\vartheta_{sk}\cos(2k)^2\eta - \pi\sum_{k=1}^{\infty}(2k)^3\frac{\omega_{sk}}{(2k)^2}\sin(2k)^2\eta}{\int_0^{\pi}\alpha h(\alpha,\kappa,w)d\alpha} \tag{2.9}$$

$$-\frac{\pi\sum_{k=1}^{\infty}(2k)^3\int_0^{\eta}h_{sk}(\kappa)q(\kappa)d\kappa}{\int_0^{\pi}\alpha h(\alpha,\kappa,w)d\alpha}$$

Definition 2.1 $w = \max_{0\le\eta\le T}\frac{|w_0(\eta)|}{2}+\sum_{k=1}^{\infty}\left(\max_{0\le\eta\le T}|w_{ck}(\eta)|+\max_{0\le\eta\le T}|w_{sk}(\eta)|\right)$ is called the Banach norm.

Theorem 2.1 If (L1)–(L3) are provided, problem (2.4)–(2.7) has a unique solution.

Proof. If we give iterations to (8), we get

$$w_0^{(N+1)} = w_0^{(0)} + \frac{2}{\pi}\int_0^{\eta}\int_0^{\pi}(\eta-\kappa)h\left(\alpha,\kappa,w^{(N)}\right)q^{(N)}(\kappa)d\alpha d\kappa$$

$$w_{ck}^{(N+1)} = w_{ck}^{(0)} + \frac{2}{4\pi k^2}\int_0^{\eta}\int_0^{\pi}h\left(\alpha,\kappa,w^{(N)}\right)q^{(N)}(\kappa)A_{\eta}\cos 2k\alpha d\alpha d\kappa$$

$$w_{sk}^{(N+1)} = w_{sk}^{(0)} + \frac{2}{4\pi k^2}\int_0^{\eta}\int_0^{\pi}h\left(\alpha,\kappa,w^{(N)}\right)q^{(N)}(\kappa)A_{\eta}\sin 2k\alpha d\alpha d\kappa$$

$$w_0^{(0)} = \vartheta_0 + \omega_0\eta$$

$$w_{ck}^{(0)} = \vartheta_{ck}\cos(2k)^2\eta + \frac{\omega_{ck}}{(2k)^2}\sin(2k)^2\eta$$

$$w_{sk}^{(0)} = \vartheta_{sk}\cos(2k)^2\eta + \frac{\omega_{sk}}{(2k)^2}\sin(2k)^2\eta$$

$$q^{(N+1)} = \frac{S''(\eta)-\pi\sum_{k=1}^{\infty}(2k)^3\vartheta_{sk}\cos(2k)^2\eta-\pi\sum_{k=1}^{\infty}(2k)^3\frac{\omega_{sk}}{(2k)^2}\sin(2k)^2\eta}{\int_0^{\pi}\alpha h\left(\alpha,\kappa,w^{(N)}\right)d\alpha}$$

$$-\frac{\pi\sum_{k=1}^{\infty}(2k)^3\int_0^{\eta}h_{sk}(\eta)q^{(N)}(\eta)d\eta}{\int_0^{\pi}\alpha h\left(\alpha,\kappa,w^{(N)}\right)d\alpha} \tag{2.10}$$

From the assumptions of the theorem, it's seen that $w^{(0)} \in \mathbf{B}$.

$$w_0^{(1)} = w_0^{(0)} + \frac{2}{\pi}\int_0^{\eta}\int_0^{\pi}(\eta-\kappa)h\left(\alpha,\kappa,w^{(0)}\right)q^{(0)}(\kappa)d\alpha d\kappa$$

After applying the Cauchy inequality, we have

$$\left|w_0^{(1)}\right| \le \left|w_0^{(0)}\right|$$

$$+\frac{2}{\pi}\left(\int_0^{\eta}\int_0^{\pi}(\eta-\kappa)^2 d\kappa\right)^{\frac{1}{2}}\left(\int_0^{\eta}\int_0^{\pi}\left(q^{(0)}(\kappa)\left[h\left(\alpha,\kappa,w^{(0)}\right)-h(\alpha,\kappa,0)\right]\right)^2 d\alpha d\kappa\right)^{\frac{1}{2}}$$

$$+\frac{2}{\pi}\left(\int_0^{\eta}\int_0^{\pi}(\eta-\kappa)^2 d\kappa\right)^{\frac{1}{2}}\left(\int_0^{\eta}\int_0^{\pi}\left(h(\alpha,\kappa,0)q^{(0)}(\kappa)\right)^2 d\alpha d\kappa\right)^{\frac{1}{2}}$$

After applying the Lipschitz condition and obtaining the maximum of both sides of the last inequalities consecutively, we get

$$\max_{0\le\eta\le T}\left|w_0^{(1)}\right| \le |\vartheta_0| + T|\psi_0| + 2\sqrt{\frac{T^3}{3\pi}}w^{(0)}{}_B q^{(0)}{}_{C(0,T)} b_{L_2(\Omega)} + 2\sqrt{\frac{T^3}{3\pi}}q^{(0)}{}_{C(0,T)} h_{L_2(\Omega)} \tag{2.11}$$

$$w_{ck}^{(1)} = w_{ck}^{(0)} + \frac{2}{4\pi k^2}\int_0^{\eta}\int_0^{\pi} h\left(\alpha,\kappa,w^{(0)}\right)q^{(0)}(\kappa)A_\eta \cos 2k\alpha d\alpha d\kappa$$

After applying the Cauchy inequality, we have

$$\left|w_{ck}^{(1)}\right| \le \left|w_{ck}^{(0)}\right|$$

$$+\frac{2}{4\pi k^2}\left(\int_0^{\eta}\int_0^{\pi} d\kappa\right)^{\frac{1}{2}}\left(\int_0^{\eta}\int_0^{\pi} h\left(\alpha,\kappa,w^{(0)}\right)q^{(0)}(\kappa)A_\eta \cos 2k\alpha d\alpha d\kappa\right)^{\frac{1}{2}}$$

$$+\frac{2}{4\pi k^2}\left(\int_0^{\eta}\int_0^{\pi} d\kappa\right)^{\frac{1}{2}}\left(\int_0^{\eta}\int_0^{\pi} h(\alpha,\kappa,0)q^{(0)}(\kappa)A_\eta \cos 2k\alpha d\alpha d\kappa\right)^{\frac{1}{2}}$$

$$-\frac{2}{4\pi k^2}\left(\int_0^{\eta}\int_0^{\pi} d\kappa\right)^{\frac{1}{2}}\left(\int_0^{\eta}\int_0^{\pi} h(\alpha,\kappa,0)q^{(0)}(\kappa)A_{\eta}\cos 2k\alpha d\alpha d\kappa\right)^{\frac{1}{2}}$$

If we take the sum of both sides $\left(k=\overline{1,\infty}\right)$ the last equations, we have

$$\left|w_{ck}^{(1)}\right| \le \sum_{k=1}^{\infty}\left|w_{ck}^{(0)}\right| + \frac{\sqrt{T}}{2\pi}\sum_{k=1}^{\infty}\frac{1}{k^2}\left(\int_0^{\eta}\int_0^{\pi}\left(q^{(0)}(\eta)\left[h\left(\alpha,\kappa,w^{(0)}\right)-h(\alpha,\kappa,0)\right]\right)^2 d\alpha d\kappa\right)^{\frac{1}{2}}$$

$$+\frac{\sqrt{T}}{2\pi}\sum_{k=1}^{\infty}\frac{1}{k^2}\left(\int_0^{\eta}\int_0^{\pi}\left(q^{(0)}(\eta)h(\alpha,\kappa,0)\right)^2 d\alpha d\kappa\right)^{\frac{1}{2}}$$

Applying Hölder inequalities, we obtain

$$\left|w_{ck}^{(1)}\right| \le \sum_{k=1}^{\infty}\left|w_{ck}^{(0)}\right| + \frac{\sqrt{T}}{2\pi}\left(\sum_{k=1}^{\infty}\frac{1}{k^2}\right)^{\frac{1}{2}}\left(\sum_{k=1}^{\infty}\int_0^{\eta}\int_0^{\pi}\left(q^{(0)}(\eta)\left[h\left(\alpha,\kappa,w^{(0)}\right)-h(\alpha,\kappa,0)\right]\right)^2 d\alpha d\eta\right)^{\frac{1}{2}}$$

$$+\frac{\sqrt{T}}{2\pi}\left(\sum_{k=1}^{\infty}\frac{1}{k^2}\right)^{\frac{1}{2}}\left(\sum_{k=1}^{\infty}\int_0^{\eta}\int_0^{\pi}\left(q^{(0)}(\eta)h(\alpha,\kappa,0)\right)^2 d\alpha d\kappa\right)^{\frac{1}{2}}$$

where $\left(\sum_{k=1}^{\infty}\frac{1}{k^2}\right)^{\frac{1}{2}} = \frac{\pi^2}{6}$. Applying the Bessel inequality and Lipschitz condition consecutively, we obtain

$$\sum_{k=1}^{\infty}\max_{0\le\eta\le T}\left|w_{ck}^{(1)}\right| \le \sum_{k=1}^{\infty}\left|\vartheta_{ck}\right| + \frac{\pi^2}{24}\sum_{k=1}^{\infty}\left|\omega_{ck}\right|$$
$$+\frac{\pi\sqrt{T}}{12}q^{(0)}{}_{C(0,T)}w^{(0)}{}_{B}b_{L_2(\Omega)}$$
$$+\frac{\pi\sqrt{T}}{12}q^{(0)}{}_{C(0,T)}h_{L_2(\Omega)}$$

and by following the same approaches, we have

$$\sum_{k=1}^{\infty}\max_{0\le\eta\le T}\left|w_{sk}^{(1)}\right| \le \sum_{k=1}^{\infty}\left|\vartheta_{sk}\right| + \frac{\pi^2}{24}\sum_{k=1}^{\infty}\left|\omega_{sk}\right|$$

$$+\frac{\pi\sqrt{T}}{12}q^{(0)}{}_{C(0,T)}w^{(0)}{}_{B}b_{L_2(\Omega)}$$

$$+\frac{\pi\sqrt{T}}{12}q^{(0)}{}_{C(0,T)}h_{L_2(\Omega)}$$

$$w^{(1)}{}_{\mathbf{B}}=\max_{0\le\eta\le T}\frac{\left|w_0^{(0)}\right|}{2}+\sum_{k=1}^{\infty}\left(\max_{0\le\eta\le T}\left|w_{ck}^{(0)}\right|+\max_{0\le\eta\le T}\left|w_{sk}^{(0)}\right|\right)$$

$$\le\frac{|\vartheta_0|}{2}+\sum_{k=1}^{\infty}\left(|\vartheta_{ck}|+|\vartheta_{sk}|\right)+\frac{\pi^2}{24}\sum_{k=1}^{\infty}\left(|\omega_{ck}|+|\omega_{sk}|\right)$$

$$+\sqrt{T}\left(\frac{12T+\pi\sqrt{3\pi}}{6\sqrt{3\pi}}\right)q^{(0)}{}_{C(0,T)}w^{(0)}{}_{B}b_{L_2(\Omega)}$$

$$+\sqrt{T}\left(\frac{12T+\pi\sqrt{3\pi}}{6\sqrt{3\pi}}\right)q^{(0)}{}_{C(0,T)}h_{L_2(\Omega)}$$

where $L=\sqrt{T}\left(\frac{12T+\pi\sqrt{3\pi}}{6\sqrt{3\pi}}\right)$. Then we have $w^{(1)}\in\mathbf{B}$. For step N,

$$w^{(N+1)}{}_{\mathbf{B}}=\max_{0\le\eta\le T}\frac{\left|w_0^{(N)}\right|}{2}+\sum_{k=1}^{\infty}\left(\max_{0\le\eta\le T}\left|w_{ck}^{(N)}\right|+\max_{0\le\eta\le T}\left|w_{sk}^{(N)}\right|\right)$$

$$\le\frac{|\vartheta_0|}{2}+\sum_{k-1}^{\infty}\left(|\vartheta_{ck}|+|\vartheta_{sk}|\right)+\frac{\pi^2}{24}\sum_{k-1}^{\infty}\left(|\omega_{ck}|+|\omega_{sk}|\right)$$

$$+Lw^{(N)}{}_{B}q^{(N)}{}_{C[0,T]}b_{L_2(\Omega)}$$

$$+Lq^{(N)}{}_{C[0,T]}h_{L_2(\Omega)}$$

From $w^{(N)}\in\mathbf{B}$ and the previous assumptions, we obtain $w^{(N+1)}\in\mathbf{B}$,

$$\{w(\eta)\}=\{w_0(\eta),w_{ck}(\eta),w_{sk}(\eta),k=1,2,\ldots\}\in\mathbf{B}.$$

$$q^{(1)}=\frac{S''(\eta)-\pi\sum_{k=1}^{\infty}(2k)^3\vartheta_{sk}\cos(2k)^2\eta-\pi\sum_{k=1}^{\infty}(2k)^3\frac{\omega_{ck}}{(2k)^2}\sin(2k)^2\eta}{\int_0^{\pi}\alpha h\left(\alpha,\kappa,w^{(0)}\right)d\alpha}$$

$$-\frac{\pi\sum_{k=1}^{\infty}(2k)^3\frac{1}{(2k)^2}\int_0^{\eta}f_{sk}(\eta)q^{(0)}(\eta)d\eta}{\int_0^{\pi}\alpha h\left(\alpha,\kappa,w^{(0)}\right)d\alpha}$$

After applying the Cauchy inequality, we have

$$\left|q^{(1)}\right| \le \frac{\left|S(\eta)\right|}{\frac{\pi^2}{2}M} + \frac{\pi\sum_{k=1}^{\infty}\left(\left|\vartheta_{ck}\right| + \left|\omega_{ck}\right|\right)}{\frac{\pi^2}{2}M} + \frac{\left(\int_0^{\eta} d\kappa\right)^{\frac{1}{2}}\left(\int_0^{\eta}\int_0^{\pi}\left(q^{(0)}\left[h\left(\alpha,\kappa,w^{(0)}\right) - h\left(\alpha,\kappa,0\right)\right]\right)^2 d\alpha d\kappa\right)^{\frac{1}{2}}}{\frac{\pi^2}{2}M} + \frac{\left(\int_0^{\eta} d\kappa\right)^{\frac{1}{2}}\left(\int_0^{\eta}\int_0^{\pi}\left(q^{(0)}h\left(\alpha,\kappa,0\right)\right)^2 d\alpha d\kappa\right)^{\frac{1}{2}}}{\frac{\pi^2}{2}M}$$

After applying the Lipschitz condition, we get

$$q^{(1)} \le \frac{2S(\eta)}{\pi^2 M} + \frac{2\sum_{k=1}^{\infty}\left(\vartheta_{ck} + \psi_{ck}\right)}{\pi^2 M} + \frac{2\sqrt{T}}{\pi M} w^{(0)}{}_{B} q^{(0)}{}_{C[0,T]} b_{L_2(\Omega)} + \frac{2\sqrt{T}}{\pi} q^{(0)}{}_{C[0,T]}$$

From the assumptions of the theorem, we have $q^{(1)} \in \mathbf{B}$. For N,

$$q^{(N+1)}(t) \le \frac{2S(\eta)}{\pi^2 M} + \frac{2\sum_{k=1}^{\infty}\left(\vartheta_{ck} + \omega_{ck}\right)}{\pi^2 M} + \frac{2\sqrt{T}}{\pi M} w^{(N)}{}_{B} q^{(N)}{}_{C[0,T]} b_{L_2(\Omega)} + \frac{2\sqrt{T}}{\pi} q^{(N)}{}_{C[0,T]}$$

From $q^{(N)} \in C[0,T]$ and the assumptions of the previous theorem, we get $q^{(N+1)} \in C[0,T]$.

Let's prove that $w^{(N+1)}, q^{(N+1)}$ is convergent for $N \to \infty$:

$$w^{(1)} - w^{(0)} = \frac{2}{\pi}\int_0^\eta \int_0^\pi (\eta-\kappa)\left[h\left(\alpha,\kappa,w^{(0)}\right) - h(\alpha,\kappa,0)\right] q^{(0)} d\alpha d\kappa$$

$$+\frac{2}{\pi}\int_0^\eta \int_0^\pi (\eta-\kappa) h(\alpha,\kappa,0) q^{(0)} d\alpha d\kappa$$

$$+\sum_{k=1}^{\infty} \cos 2kx \left[\frac{2}{4\pi k^2}\int_0^\eta \int_0^\pi \left[h\left(\alpha,\kappa,w^{(0)}\right) - h(\alpha,\kappa,0)\right] q^{(0)} \Lambda_\eta \cos 2k\alpha d\alpha d\kappa\right]$$

$$+\sum_{k=1}^{\infty} \cos 2kx \left[\frac{2}{4\pi k^2}\int_0^\eta \int_0^\pi h(\alpha,\kappa,0) q^{(0)} \Lambda_\eta \cos 2k\alpha d\alpha d\kappa\right]$$

$$+\sum_{k=1}^{\infty} \sin 2kx \left[\frac{2}{4\pi k^2}\int_0^\eta \int_0^\pi \left[h\left(\alpha,\kappa,w^{(0)}\right) - h(\alpha,\kappa,0)\right] q^{(0)} \Lambda_\eta \sin 2k\alpha d\alpha d\kappa\right]$$

$$+\sum_{k=1}^{\infty} \sin 2kx \left[\frac{2}{4\pi k^2}\int_0^\eta \int_0^\pi h(\alpha,\kappa,0) q^{(0)} \Lambda_\eta \sin 2k\alpha d\alpha d\kappa\right]$$

After applying the Cauchy inequality, we have

$$\left|w^{(1)} - w^{(0)}\right| \le \frac{2}{\pi}\left(\int_0^\eta \int_0^\pi (\eta-\kappa)^2 d\kappa\right)^{\frac{1}{2}} \left(\int_0^\eta \int_0^\pi \left(\left[h\left(\alpha,\kappa,w^{(0)}\right) - h(\alpha,\kappa,0)\right] q^{(0)}\right)^2 d\alpha d\kappa\right)^{\frac{1}{2}}$$

$$+\frac{2}{\pi}\left(\int_0^\eta \int_0^\pi (\eta-\kappa)^2 d\kappa\right)^{\frac{1}{2}} \left(\int_0^\eta \int_0^\pi \left(h(\alpha,\kappa,0) q^{(0)}(\eta)\right)^2 d\alpha d\eta\right)^{\frac{1}{2}}$$

$$+\frac{2}{4\pi k^2}\left(\int_0^\eta \int_0^\pi d\kappa\right)^{\frac{1}{2}} \left(\int_0^\eta \int_0^\pi \left(\left[h\left(\alpha,\kappa,w^{(0)}\right) - h(\alpha,\kappa,0)\right] q^{(0)} \cos 2k\alpha\right)^2 d\alpha d\kappa\right)^{\frac{1}{2}}$$

$$+\frac{2}{4\pi k^2}\left(\int_0^\eta \int_0^\pi d\kappa\right)^{\frac{1}{2}} \left(\int_0^\eta \int_0^\pi \left(h(\alpha,\kappa,0) q^{(0)} \cos 2k\alpha\right)^2 d\alpha d\kappa\right)^{\frac{1}{2}}$$

$$+\frac{2}{4\pi k^2}\left(\int_0^\eta\int_0^\pi d\kappa\right)^{\frac{1}{2}}\left(\int_0^\eta\int_0^\pi\left(\left[h\left(\alpha,\kappa,w^{(0)}\right)-h\left(\alpha,\kappa,0\right)\right]q^{(0)}\sin 2k\alpha\right)^2 d\alpha d\kappa\right)^{\frac{1}{2}}$$

$$+\frac{2}{4\pi k^2}\left(\int_0^\eta\int_0^\pi d\kappa\right)^{\frac{1}{2}}\left(\int_0^\eta\int_0^\pi\left(h\left(\alpha,\kappa,0\right)q^{(0)}\sin 2k\alpha\right)^2 d\alpha d\kappa\right)^{\frac{1}{2}}$$

After applying the Hölder, Bessel inequality and Lipschitz condition consecutively, we obtain

$$w^{(1)}-w^{(0)}{}_B \le Lw^{(0)}{}_B q^{(0)}{}_{C[0,T]} b_{L_2(\Omega)}$$

$$+Lq^{(0)}{}_{C[0,T]} h_{L_2(\Omega)}$$

$$A = Lw^{(0)}{}_B q^{(0)}{}_{C[0,T]} b_{L_2(\Omega)}$$

$$+Lq^{(0)}{}_{C[0,T]} h_{L_2(\Omega)}$$

$$q^{(1)}-q^{(0)}{}_{C[0,T]} \le \frac{2\sqrt{T}}{\pi MC} w^{(1)}-w^{(0)}{}_B q^{(0)}{}_{C[0,T]} b_{L_2(\Omega)}$$

where $C_1=\left(1-\frac{2\sqrt{T}}{\pi M}w^{(0)}{}_B\right)$

$$w^{(2)}-w^{(1)}{}_B$$

$$\le Lw^{(1)}-w^{(0)}{}_B q^{(1)}{}_{C[0,T]}$$

$$+Lq^{(1)}-q^{(0)}{}_{C[0,T]} M$$

$$w^{(2)}-w^{(1)}{}_B$$

$$\le Lw^{(1)}-w^{(0)}{}_B q^{(1)}{}_{C[0,T]} b_{L_2(\Omega)}$$

$$+L\frac{2\sqrt{T}}{\pi C_1} w^{(1)}-w^{(0)}{}_B q^{(0)}{}_{C[0,T]} b_{L_2(\Omega)}$$

$$w^{(2)}-w^{(1)}{}_B \le LAD_1 b_{L_2(\Omega)}$$

where $D_1=\left(q^{(1)}{}_{C[0,T]}+\frac{2\sqrt{T}}{\pi C}q^{(0)}{}_{C[0,T]}\right)$

$$q^{(2)}-q^{(1)}{}_{C[0,T]} \le \frac{2\sqrt{T}}{\pi MC_2} w^{(2)}-w^{(1)}{}_B q^{(2)}{}_{C[0,T]} b_{L_2(\Omega)}$$

where $C_2 = \left(1 - \frac{2\sqrt{T}}{\pi M} w^{(1)}{}_B\right)$

For N, we get

$$q^{(N+1)} - q^{(N)}{}_{C[0,T]} \leq \frac{2\sqrt{T}}{\pi M C_N} w^{(N+1)} - w^{(N)}{}_B q^{(N+1)}{}_{C[0,T]} b_{L_2(\Omega)}$$

where $C_N = \left(1 - \frac{2\sqrt{T}}{\pi M} w^{(N-1)}{}_B\right)$

$$w^{(N+1)} - w^{(N)}{}_B \leq \frac{L^N A D_N}{\sqrt{N!}} b^N_{L_2(\Omega)}$$

$$w^{(N+1)} \to w^{(N)},\ N \to \infty, \text{ then } q^{(N+1)} \to q^{(N)},\ N \to \infty.$$

where $D_N = \left(q^{(N)}{}_{C[0,T]} + \frac{2\sqrt{T}}{\pi C} q^{(N-1)}{}_{C[0,T]}\right)$. Let it show

$$\lim_{N\to\infty} w^{(N+1)} = w,\ \lim_{N\to\infty} q^{(N+1)} = q$$

$$w - w^{(N+1)} = \frac{2}{\pi}\int_0^\eta\int_0^\pi (\eta-\kappa)\left[h(\alpha,\kappa,w) - h\left(\alpha,\kappa,w^{(N+1)}\right)\right] q\,d\alpha\,d\kappa$$

$$+\frac{2}{\pi}\int_0^\eta\int_0^\pi (\eta-\kappa)\left[h\left(\alpha,\kappa,w^{(N+1)}\right) - h\left(\alpha,\kappa,w^{(N)}\right)\right] q\,d\alpha\,d\kappa$$

$$+\frac{2}{\pi}\int_0^\eta\int_0^\pi (\eta-\kappa)\,h\left(\alpha,\kappa,w^{(N)}\right)\left[q - q^{(N)})\right] d\alpha\,d\kappa$$

$$+\sum_{k=1}^{\infty}\cos 2kx\left[\frac{2}{4\pi k^2}\int_0^\eta\int_0^\pi\left[h(\alpha,\kappa,w) - h\left(\alpha,\kappa,w^{(N+1)}\right)\right] qA_\eta \cos 2k\alpha\,d\alpha\,d\kappa\right]$$

$$+\sum_{k=1}^{\infty}\cos 2kx\left[\frac{2}{4\pi k^2}\int_0^\eta\int_0^\pi\left[h\left(\alpha,\kappa,w^{(N+1)}\right) - h\left(\alpha,\kappa,w^{(N)}\right)\right] qA_\eta \cos 2k\alpha\,d\alpha\,d\kappa\right]$$

$$+\sum_{k=1}^{\infty}\cos 2kx\left[\frac{2}{4\pi k^2}\int_0^\eta\int_0^\pi h\left(\alpha,\kappa,w^{(N)}\right)\left[q - q^{(N)}\right]\Lambda_\eta \cos 2k\alpha\,d\alpha\,d\kappa\right]$$

$$+\sum_{k=1}^{\infty}\sin 2kx\left[\frac{2}{4\pi k^2}\int_0^\eta\int_0^\pi\left[h(\alpha,\kappa,w) - h\left(\alpha,\kappa,w^{(N+1)}\right)\right] qA_\eta \sin 2k\alpha\,d\alpha\,d\kappa\right]$$

$$+\sum_{k=1}^{\infty}\sin 2kx\left[\frac{2}{4\pi k^2}\int_0^{\eta}\int_0^{\pi}\left[h\left(\alpha,\kappa,w^{(N+1)}\right)-h\left(\alpha,\kappa,w^{(N)}\right)\right]qA_{\eta}\sin 2k\alpha d\alpha d\kappa\right]$$

$$+\sum_{k=1}^{\infty}\sin 2kx\left[\frac{2}{4\pi k^2}\int_0^{\eta}\int_0^{\pi}h\left(\alpha,\kappa,w^{(N)}\right)\left[q-q^{(N)}\right]\Lambda_{\eta}\sin 2k\alpha d\alpha d\kappa\right]$$

After using the Cauchy, Hölder inequalities consecutively, we obtain

$$\left|w-w^{(N+1)}\right| \leq 2\sqrt{\frac{T^3}{3\pi}}\left(\int_0^{\eta}\int_0^{\pi}\left(\left[h(\alpha,\kappa,w)-h\left(\alpha,\kappa,w^{(N+1)}\right)\right]qd\alpha d\kappa\right)^2\right)^{\frac{1}{2}}$$

$$2\sqrt{\frac{T^3}{3\pi}}\left(\int_0^{\eta}\int_0^{\pi}\left(\left[h\left(\alpha,\kappa,w^{(N+1)}\right)-h\left(\alpha,\kappa,w^{(N)}\right)\right]qd\alpha d\kappa\right)^2\right)^{\frac{1}{2}}$$

$$+2\sqrt{\frac{T^3}{3\pi}}\left(\int_0^{\eta}\int_0^{\pi}\left(h\left(\alpha,\kappa,w^{(N)}\right)\left[q-q^{(N)}\right]d\alpha d\kappa\right)^2\right)^{\frac{1}{2}}$$

$$+\frac{\sqrt{T}}{2\pi}\left(\sum_{k=1}^{\infty}\frac{1}{k^2}\right)^{\frac{1}{2}}\left(\sum_{k=1}^{\infty}\int_0^{\eta}\int_0^{\pi}\left(\left[h(\alpha,\kappa,w)-h\left(\alpha,\kappa,w^{(N+1)}\right)\right]qd\alpha d\kappa\right)^2\right)^{\frac{1}{2}}$$

$$+\frac{\sqrt{T}}{2\pi}\left(\sum_{k=1}^{\infty}\frac{1}{k^2}\right)^{\frac{1}{2}}\left(\sum_{k=1}^{\infty}\int_0^{\eta}\int_0^{\pi}\left(\left[h\left(\alpha,\kappa,w^{(N+1)}\right)-h\left(\alpha,\kappa,w^{(N)}\right)\right]q\cos 2k\alpha d\alpha d\kappa\right)^2\right)^{\frac{1}{2}}$$

$$+\frac{\sqrt{T}}{2\pi}\left(\sum_{k=1}^{\infty}\frac{1}{k^2}\right)^{\frac{1}{2}}\left(\sum_{k=1}^{\infty}\int_0^{\eta}\int_0^{\pi}\left(h\left(\alpha,\kappa,w^{(N)}\right)\left[q-q^{(N)}\right]\cos 2k\alpha d\alpha d\kappa\right)^2\right)^{\frac{1}{2}}$$

$$+\frac{\sqrt{T}}{2\pi}\left(\sum_{k=1}^{\infty}\frac{1}{k^2}\right)^{\frac{1}{2}}\left(\sum_{k=1}^{\infty}\int_0^{\eta}\int_0^{\pi}\left(\left[h(\alpha,\kappa,w)-h\left(\alpha,\kappa,w^{(N+1)}\right)\right]q\sin 2k\alpha d\alpha d\kappa\right)^2\right)^{\frac{1}{2}}$$

$$+\frac{\sqrt{T}}{2\pi}\left(\sum_{k=1}^{\infty}\frac{1}{k^2}\right)^{\frac{1}{2}}\left(\sum_{k=1}^{\infty}\int_0^{\eta}\int_0^{\pi}\left(\left[h\left(\alpha,\kappa,w^{(N+1)}\right)-h\left(\alpha,\kappa,w^{(N)}\right)\right]q\sin 2k\alpha d\alpha d\kappa\right)^2\right)^{\frac{1}{2}}$$

$$+\frac{\sqrt{T}}{2\pi}\left(\sum_{k=1}^{\infty}\frac{1}{k^2}\right)^{\frac{1}{2}}\left(\sum_{k=1}^{\infty}\int_0^{\eta}\int_0^{\pi}\left(h\left(\alpha,\kappa,w^{(N)}\right)\left[q-q^{(N)}\right]\sin 2k\alpha d\alpha d\kappa\right)^2\right)^{\frac{1}{2}}$$

After applying the Bessel inequality and Lipschitz condition consecutively, we obtain

$$\begin{aligned} & w-w^{(N+1)}{}_B \\ & \leq Lw-w^{(N+1)}{}_B q_{C[0,T]} b_{L_2(\Omega)} \\ & +Lw^{(N+1)}-w^{(N)}{}_B q_{C[0,T]} b_{L_2(\Omega)} \\ & +Lq(t)-q^{(N+1)}{}_{C[0,T]} M \end{aligned} \tag{2.12}$$

$$q-q^{(N+1)}{}_{C[0,T]} \leq \frac{2\sqrt{T}}{\pi M C_N} w-w^{(N+1)}{}_B q^{(N+1)}{}_{C[0,T]} b_{L_2(\Omega)}$$

Using Gronwall's inequality for the last equations, we yield

$$\begin{aligned} & w-w^{(N+1)2}{}_B \\ & \left(2\left[\frac{L^N AD_N}{\sqrt{N!}} b^{N+1}{}_{L_2(\Omega)}\right]^2+K\right) \\ & \times \exp 2L^2 q^2_{C[0,T]} b^2_{L_2(\Omega)} \end{aligned} \tag{2.13}$$

where

$$K=\frac{2\sqrt{T}}{\pi C_N} Lq^{(N+1)}{}_{C[0,T]} b_{L_2(\Omega)}$$

$$w^{(N+1)} \rightarrow w, q^{(N+1)} \rightarrow q,\ N \rightarrow \infty$$

For uniqueness, let (w,q), (v,h) be two solutions of (2.4)–(2.7). The same operations for $|w-v|$ and $|q-h|$ obtain

$$\begin{aligned} w-v_B \leq\ & Lq-h_{C[0,T]} w_B \\ & +L\left(\int_0^{\eta}\int_0^{\pi} b^2 |w-v|^2\, d\alpha d\kappa\right)^{\frac{1}{2}} \end{aligned}$$

$$q-h_{C[0,T]} \leq \frac{2\sqrt{T}}{\pi MC} w-v_B q_{C[0,T]} b_{L_2(\Omega)}$$

$$w - v_B \leq L\left(\int_0^{\eta}\int_0^{\pi} b^2 \left|w - v\right|^2 d\alpha d\kappa\right)^{\frac{1}{2}} \tag{2.14}$$

If the inequalities in (2.13) are applied to Gronwall's inequality, then we get $w = v$. Hence we have $q = h$. The proof is completed.

2.3 STABILITY OF THE SOLUTION RELATIVE UPON THE DATA

Theorem 2.2 *According to (L1)–(L3), the solution of (1)–(4) is constantly dependent on the data.*

Proof. Suppose $\Psi = \left\{\vartheta, \omega, S, h\right\}$, $\overline{\Psi} = \left\{\overline{\vartheta}, \overline{\omega}, \overline{S}, h\right\}$, M_i, $i = 1,2$ such that

$$S_{C^1[0,T]} \leq M_1, \overline{S}_{C^1[0,T]} \leq M_1$$

$$\vartheta_{C^3[0,\pi]} \leq M_2, \overline{\vartheta}_{C^3[0,\pi]} \leq M_2$$

$$\omega_{C^3[0,\pi]} \leq M_2, \overline{\omega}_{C^3[0,\pi]} \leq M_2$$

Let $\Psi = \left(S_{C^1[0,T]} + \vartheta_{C^3[0,\pi]} + \omega_{C^1[0,\pi]} + h_{C^{3,0}(\overline{\Gamma})}\right)$. Let (g, w) and $(\overline{g}, \overline{w})$ be solutions.

$$w - \overline{w} = \frac{\left(\vartheta_0 - \overline{\vartheta_0}\right)}{2} + \frac{\left(\omega_0 - \overline{\omega_0}\right)\eta}{2} + \sum_{k=1}^{\infty} \cos 2kx\left(\vartheta_{ck} - \overline{\vartheta_{ck}}\right)\cos(2k)^2 \eta$$

$$+ \sum_{k=1}^{\infty} \sin 2kx\left(\vartheta_{sk} - \overline{\vartheta_{sk}}\right)\sin(2k)^2 \eta$$

$$\sum_{k=1}^{\infty} \cos 2kx \frac{\left(\omega_{ck} - \overline{\omega_{ck}}\right)}{(2k)^2}\cos(2k)^2 \eta + \sum_{k=1}^{\infty} \sin 2kx \frac{\left(\omega_{sk} - \overline{\omega_{sk}}\right)}{(2k)^2}\sin(2k)^2 \eta$$

$$+ \frac{2}{\pi}\int_0^{\eta}\int_0^{\pi} (\eta - \kappa)\left[h(\alpha, \kappa, w) - h(\alpha, \kappa, \overline{w})\right] q \, d\alpha d\kappa$$

$$+ \frac{2}{\pi}\int_0^{\eta}\int_0^{\pi} (\eta - \kappa) h(\alpha, \kappa, \overline{w})\left[q - \overline{q}\right] d\alpha d\kappa$$

$$+ \sum_{k=1}^{\infty} \cos 2kx\left|\frac{2}{4\pi k^2}\int_0^{\eta}\int_0^{\pi}\left[h(\alpha, \kappa, w) - h(\alpha, \kappa, \overline{w})\right] q \Lambda_{\eta} \cos 2k\alpha \, d\alpha d\kappa\right|$$

$$+ \sum_{k=1}^{\infty} \cos 2kx\left|\frac{2}{4\pi k^2}\int_0^{\eta}\int_0^{\pi} h(\alpha, \kappa, \overline{w})\left[q - \overline{q}\right] \Lambda_{\eta} \cos 2k\alpha \, d\alpha d\kappa\right|$$

$$+\sum_{k=1}^{\infty}\sin 2kx\left[\frac{2}{4\pi k^2}\int_0^{\eta}\int_0^{\pi}\left[h(\alpha,\kappa,w)-h(\alpha,\kappa,\overline{w})\right]q\Lambda_{\eta}\sin 2k\alpha d\alpha d\kappa\right]$$

$$+\sum_{k=1}^{\infty}\sin 2kx\left[\frac{2}{4\pi k^2}\int_0^{\eta}\int_0^{\pi}h(\alpha,\kappa,\overline{w})\left[q-\overline{q}\right]\Lambda_{\eta}\sin 2k\alpha d\alpha d\kappa\right]$$

$$w-\overline{w}_B \leq \vartheta-\overline{\vartheta}+\psi-\overline{\psi}\left|T\right|+Lw-\overline{w}_B q_{C[0,T]} b_{L_2(\Omega)}$$

$$+Lq-\overline{q}_{C[0,T]}M$$

$$q-\overline{q}_{C[0,T]} \leq \frac{2\sqrt{T}}{\pi MC}w-\overline{w}_B q_{C[0,T]} b_{L_2(\Omega)}$$

$$w-\overline{w}_B \leq \vartheta-\overline{\vartheta}+\omega-\overline{\omega}\left|T\right|+Lw-\overline{w}_B q_{C[0,T]} b_{L_2(\Omega)}$$

$$+L\frac{2\sqrt{T}}{\pi C}w-\overline{w}_B q_{C[0,T]} b_{L_2(\Omega)}$$

$$w-\overline{w}_B \leq \vartheta-\overline{\vartheta}+\omega-\overline{\omega}\left|T\right|+L\left(1+\frac{2\sqrt{T}}{\pi}\right)w-\overline{w}_B q_{C[0,T]} b_{L_2(\Omega)} \tag{2.15}$$

$$w-\overline{w}_B^2 \leq 2M_3^2\Psi-\overline{\Psi}^2$$

$$\times\exp 2M_4^2\left(\int_0^{\eta}\int_0^{\pi}q^2b^2 d\alpha d\kappa\right)$$

where

$$M_3 = \max\left\{1,\left|T\right|\right\}$$

$$M_4 = L\left(1+\frac{2\sqrt{T}}{\pi}\right)$$

For $\Psi \to \overline{\Psi}$, $w \to \overline{w}$ than $q \to \overline{q}$.

2.4 ANALYSIS OF NUMERICAL METHOD

After linearization of the nonlinear terms of (2.3)–(2.5), the following problem is obtained:

$$\frac{\partial^2 w^{(n)}}{\partial \eta^2}+\frac{\partial^4 w^{(n)}}{\partial x^4}=q(\eta)h\left(x,\eta,w^{(n-1)}\right),(x,\eta)\in\Omega \tag{2.16}$$

$$w^{(n)}(0,\eta)=w^{(n)}(\pi,\eta)$$

$$w_x^{(n)}(0,\eta)=w_x^{(n)}(\pi,\eta)$$

$$w_{xx}^{(n)}(0,\eta) = w_{xx}^{(n)}(\pi,\eta) \tag{2.17}$$

$$w_{xxx}^{(n)}(0,\eta) = w_{xxx}^{(n)}(\pi,\eta), \eta \in [0,T]$$

$$w^{(n)}(x,0) = \vartheta(x), w_{\eta}^{(n)}(x,0) = \omega(x), x \in [0,\pi] \tag{2.18}$$

$$S(\eta) = \int_0^{\pi} \alpha w^{(n)}(\alpha,\kappa) d\alpha, \eta \in [0,T] \tag{2.19}$$

If we write $w^{(n)}(x,\eta) = u(x,\eta)$ and $h\left(x,\eta,w^{(n-1)}\right) = \tilde{h}(x,\eta)$, then the new linear problem is

$$\frac{\partial^2 u}{\partial \eta^2} + \frac{\partial^4 u}{\partial x^4} = q(\eta)\tilde{h}(x,\eta), (x,\eta) \in \Omega \tag{2.20}$$

$$u(0,\eta) = u(\pi,\eta)$$

$$u_x(0,\eta) = u_x(\pi,\eta)$$

$$u_{xx}(0,\eta) = u_{xx}(\pi,\eta) \tag{2.21}$$

$$u_{xxx}(0,\eta) = u_{xxx}(\pi,\eta), \eta \in [0,T]$$

$$u(x,0) = \vartheta(x), u_{\eta}(x,0) = \omega(x), x \in [0,\pi] \tag{2.22}$$

$$S(\eta) = \int_0^{\pi} \alpha u(\alpha,\eta) d\alpha, \eta \in [0,T] \tag{2.23}$$

If we move on to the finite difference approach to the previous problem

$$\frac{1}{\tau^2}\left(u_i^{j+1} - 2u_i^j + u_i^{j-1}\right) + \frac{1}{k^4}\left(u_{i+2}^j - 4u_{i+1}^j + 6u_i^j - 4u_{i-1}^j + u_{i-2}^j\right) = q^j \tilde{h}_i^j \tag{2.24}$$

$$u_i^0 = \phi_i, \frac{1}{\tau}\left(u_i^1 - u_i^0\right) = \psi_i \tag{2.25}$$

$$u_0^j = u_{N_x+1}^j, u_1^j = u_{N_x+2}^j \tag{2.26}$$

$$u_{-1}^j = u_{N_x}^j, u_2^j - u_{-2}^j = u_{N_x+3}^j - u_{N_x-1}^j \tag{2.27}$$

Let's convert the interval $[0,\pi] \times [0,T]$ to an interval $N_x \times N_t$, with $k = \pi / N_x$ and $\tau = T / N_t$.

Let x_i, t_j (grid point) be as follows:

$$x_i = ik; i = 0,1,2,\ldots,N_x$$

$$t_j = j\tau; \ j = 0,1,2,\ldots,N_t;$$

$$v_i^j = u(x_i,t_j), \tilde{h}_i^j = \tilde{h}(x_i,t_j), q^j = q(t_j)$$

Integrating (2.15) with respect to x from 0 to π and using (2.16) and (2.18), the function $q(\eta)$ is obtained:

$$q(\eta)=\frac{\left[S''(\eta)+\pi u_{xxx}(\pi,\eta)\right]}{\int_0^\pi \alpha\tilde{h}(\alpha,\kappa)d\alpha} \tag{2.28}$$

$q(t)$ is approximated as

$$q^j=\frac{\left[\left(\left(S^{j+1}-2S^j+S^{j-1}\right)/\tau^2\right)+\pi\left(u^j_{N_x+3}-2u^j_{N_x+2}+2u^j_{N_x}-u^j_{N_x-1}\right)\right]}{\left(\int_0^\pi \alpha\,\widetilde{fh}^j_i\,d\alpha\right)}$$

where $S^j=S(t_j)$, $j=0,1,\ldots,N_t$, by the Simpson rule.

$q^{j(s)}$, $u_i^{j(s)}$ are q^j, u_i^j at the iteration s-th step; $q^{j(s)}$ is as follows

$$q^{j(s)}=\frac{\left[\left(\left(S^{j+1(s+1)}-2S^{j(s)}+S^{j-1(s-1)}\right)/\tau^2\right)+\pi\left(u^{j(s)}_{N_x+3}-2u^{j(s)}_{N_x+2}+2u^{j(s)}_{N_x}-u^{j(s)}_{N_x-1}\right)\right]}{\left(\int_0^\pi \alpha\,\widetilde{fh}^{j(s)}_i\,d\alpha\right)}$$

The iteration of (2.23)–(2.26) is

$$\frac{1}{\tau^2}\left(u_i^{j+1(s+1)}-2u_i^{j(s)}+u_i^{j-1(s-1)}\right)+\frac{1}{k^4}\left(u_{i+2}^{j(s)}-4u_{i+1}^{j(s)}+6u_i^{j(s)}-4u_{i-1}^{j(s)}+u_{i-2}^{j(s)}\right)$$
$$=q^{j(s)}\tilde{h}_i^{j(s)} \tag{2.29}$$

$$u_i^0=\phi_i,\frac{1}{\tau}\left(u_i^1-u_i^0\right)=\psi_i \tag{2.30}$$

$$u_0^{j(s)}=u_{N_x+1}^{j(s)} \tag{2.31}$$

$$u_1^{j(s)}=u_{N_x+2}^{j(s)} \tag{2.32}$$

$$u_{-1}^{j(s)}=u_{N_x}^{j(s)} \tag{2.33}$$

$$u_2^{j(s)}-u_{-2}^{j(s)}=u_{N_x+3}^{j(s)}-u_{N_x-1}^{j(s)} \tag{2.34}$$

From the system of equations (2.28)–(2.33), $u_i^{j+1(s+1)}$ is found. If it gets the desired tolerance, the iteration is finished.

Example 2.1 The aim is to find the accurate solution

$$\{q(\eta), w(x,\eta)\} = \{\exp(2\eta), (\exp(\eta)+\exp(2\eta))\sin 2x\}$$

with the following given functions examined:

$$\varphi(x) = \sin 2x,\ E(\eta) = -\frac{\pi}{2}(\exp(\eta)+\exp(2\eta))$$

$$h(x,\eta,w) = \frac{(17\exp(\eta)+20\exp(2\eta))\sin 2x + w}{\exp(2\eta)}$$

$k = 0.0393$, $\tau = 0.005$ (*The step sizes*)

$\left| q^{k+1(s+1)} - q^{k+1(s)} \right| \le k/100$

The difference between the accurate solution and approximated solution is in Figures 2.1–2.4 when $T = 1$ as follows:

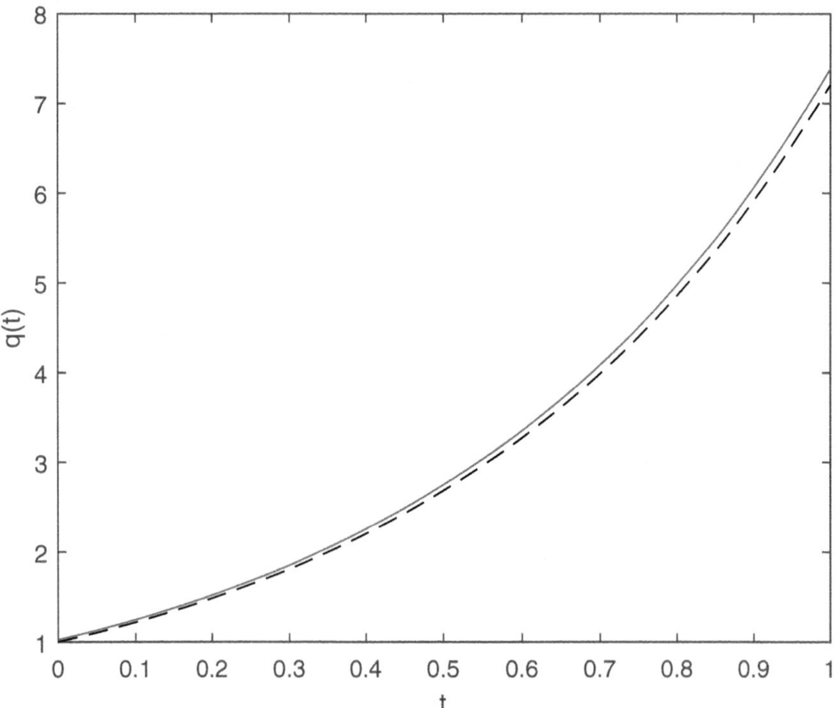

FIGURE 2.1 The accurate and approximated solutions of q. The approximated solution is indicated by the dashed black line, and the accurate solution is indicated by the red line.

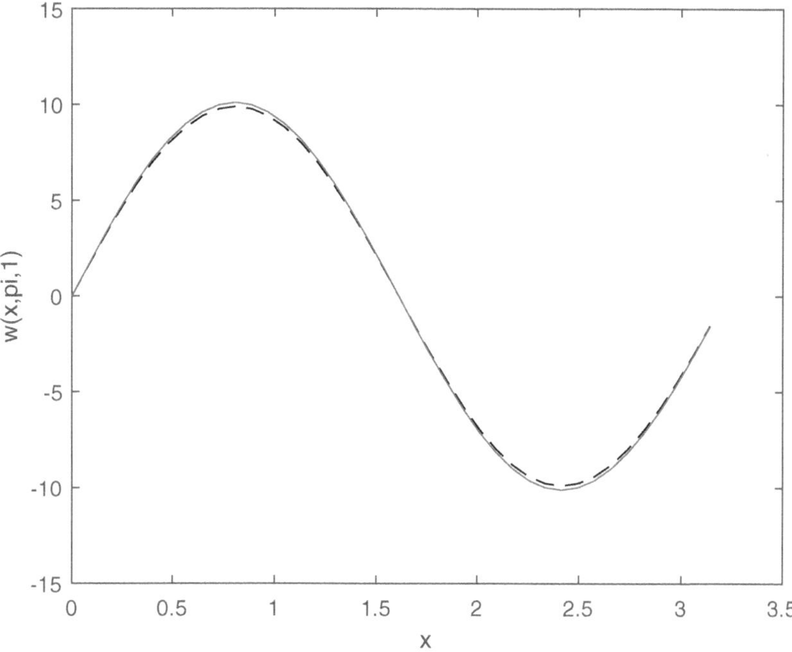

FIGURE 2.2 The accurate and approximated solutions of $w(x,pi)$. The approximated solution is indicated by the dashed black line, and the accurate solution is indicated by the red line.

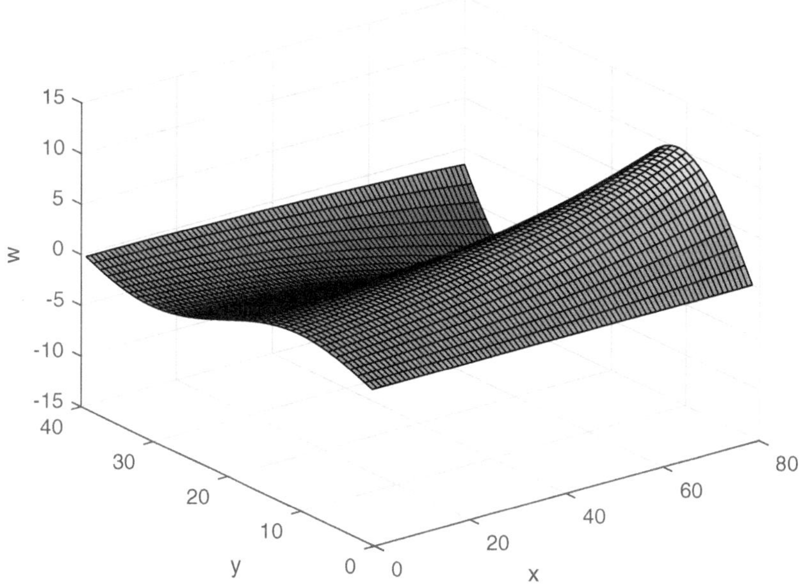

FIGURE 2.3 The accurate solutions of w.

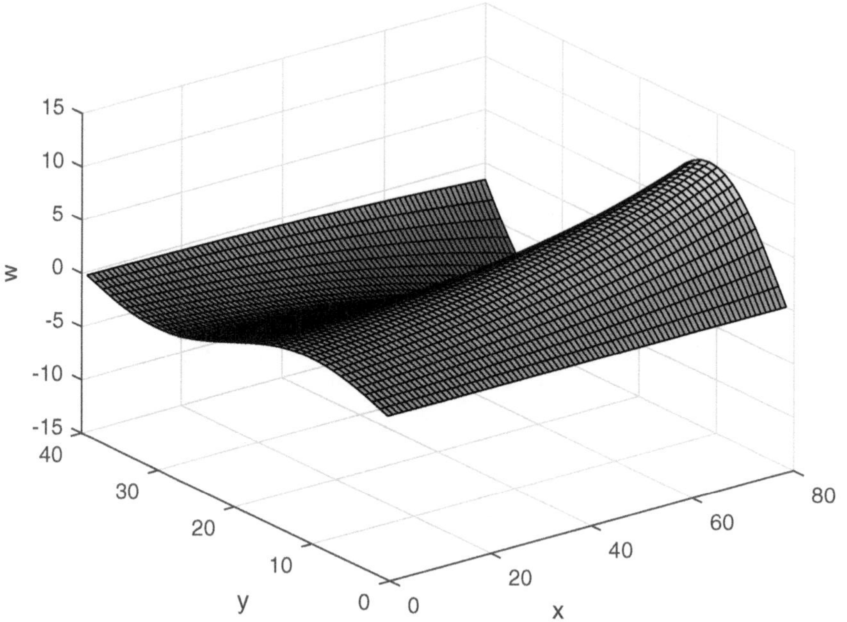

FIGURE 2.4 The approximated solutions of w.

2.5 CONCLUSION

In this study, the quasi-linear Euler Bernoulli equation was investigated using periodic and integral boundary conditions. The problem was studied both theoretically and numerically. Periodic boundary conditions were studied. Nonlocal periodic boundary conditions for heat inverse coefficient problems are more difficult than local boundary conditions. In this chapter, they were obtained by using the Fourier method and the finite-difference method.

REFERENCES

[1] H. P. W. Gottlieb, "Isospectral Euler-Bernoulli beams with continuous density and rigidity functions," *Proceedings of the Royal Society of London Series A: Mathematical, Physical and Engineering Sciences*, vol. 413, no. 1844, pp. 235–250, 1987.

[2] C. W. Soh, "Euler-Bernoulli beams from a symmetry standpoint—characterization of equivalent equations," *Journal of Mathematical Analysis and Applications*, vol. 345, no. 1, pp. 387–395, 2008.

[3] O. I. Morozov and C. W. Soh, "The equivalence problem for the Euler-Bernoulli beam equation via Cartan's method," *Journal of Physics A: Mathematical and Theoretical*, vol. 41, no. 13, 135206, pp. 135–206, 2008.

[4] J. C. Ndogmo, "Equivalence transformations of the Euler-Bernoulli equation," *Nonlinear Analysis: Real World Applications*, vol. 13, no. 5, pp. 2172–2177, 2012.

[5] E. Ozkaya and M. Pakdemirli, "Group-theoretic approach to axially accelerating beam problem," *Acta Mechanica*, vol. 155, no. 1–2, pp. 111–123, 2002.

[6] A. H. Bokhari, F. M. Mahomed, and F. D. Zaman, "Invariant boundary value problems for a fourth-order dynamic Euler-Bernoulli beam equation," *Journal of Mathematical Physics*, vol. 53, no. 4, 2012.

[7] X. Q. He, S. Kitipornchai, and K. M. Liew, "Buckling analysis of multi-walled carbon nanotubes a continuum model accounting for van der Waals interaction," *Journal of the Mechanics and Physics of Solids*, vol. 53, pp. 303–326, 2005.

[8] T. Natsuki, Q. Q. Ni, and M. Endo, "Wave propagation in single-and double-walled carbon nano tubes filled with fluids," *Journal of Applied Physics*, vol. 101, p. 034319, 2007.

[9] Y. Yana, X. Q. Heb, L. X. Zhanga, and C. M. Wang, "Dynamic behavior of triple-walled carbon nano-tubes conveying fluid," *Journal of Sound and Vibration*, vol. 319, pp. 1003–1018, 2010.

[10] T. S. Jang, "A new solution procedure for a nonlinear infinite beam equation of motion," *Communications in Nonlinear Science and Numerical Simulation*, vol. 39, pp. 321–331, 2016.

[11] T. S. Jang, "A general method for analyzing moderately large deflections of a non-uniform beam: An infinite Bernoulli-Euler–von Karman beam on a non-linear elastic foundation," *Acta Mechanica*, vol. 225, pp. 1967–1984, 2014.

[12] A. Mohebbi and M. Abbasi, "A fourth-order compact difference scheme for the parabolic inverse problem with an overspecification at a point," *Inverse Problems in Science and Engineering*, vol. 23, no. 3, pp. 457–478, 2014. http://doi.org/10.1080/17415977.2014.922075.

[13] R. Pourgholia, M. Rostamiana, and M. Emamjome, "A numerical method for solving a nonlinear inverse parabolic problem," *Inverse Problems in Science and Engineering*, vol. 18, no. 8, pp. 1151–1164, 2010.

[14] I. Baglan and F. Kanca, "An inverse coefficient problem for a quasilinear parabolic equation with periodic boundary and integral overdetermination condition," *Mathematical Methods in the Applied Sciences*, 2015. http://doi.org/10.1002/mma.3112.

[15] F. Kanca and I. Baglan, "Inverse problem for Euler-Bernoulli equation with periodic boundary condition," *Filomat*, vol. 32, no. 16, pp. 5691–5705, 2018.

[16] A. Yokus, "On the exact and numerical solutions to the FitzHugh–Nagumo equation," *International Journal of Modern Physics B*, vol. 34, no. 17, 2020.

[17] D. Kaya, A. Yokus, and U. Demiroglu, "Comparison of exact and numerical solutions for the sharma–tasso–olver equation," *Numerical Solutions of Realistic Nonlinear Phenomena*, pp. 53–65, 2020.

[18] A. Yokus, "Numerical solution for space and time fractional order Burger type equation," *Alexandria Engineering Journal*, vol. 57, no. 3, pp. 2085–2091, 2018.

[19] G. W. Hill, "On the part of the motion of the lunar perigee which is a function of the mean motions of the sun and moon," *Acta Mathematica*, vol. 8, pp. 1–36, 1986.

3 A Classification of Focal Surfaces of a Tube Surface in E^3

Sezgin Büyükkütük, İlim Kişi, and Günay Öztürk

3.1 INTRODUCTION

The curvature centers of the perpendicular section curves at a point on a regular surface correspond to a certain part of the normal vector in Euclidean $3-$ space. The maximum and minimum values of these parts are the curvature centers of the two principle curves. These two points are called focal points of the normal. In this way, focal surface is defined as the geometric locus of the principle curvature centers [1].

Thanks to their structure and properties, the focal point of the surface and focal surfaces are important concepts in terms of being subject to architecture and geometric design. In study [2], Petrusevski et al. (2017) studied the focal point of a surface and its directrix. Using these concepts, they define a focal-directional surface and apply this surface to architecture. In the study, the focal point (focus of a surface) and directrices are defined by using Cartesian coordinates. Accordingly, one can say that the surface is expressed by its directrices and focuses in the suitable global spherical coordinate system. Quadrilateral or triangular meshes are used to model the surface. Therefore, in order to obtain quadrilateral or triangular meshes, the grid points are determined on the surface. The discrete model of the surface is created in a Cartesian and spherical grid.

In a thesis [3], Petrovic (2016) discussed focal-directional surfaces and some different surfaces and gave applications to architecture (see Figure 3.2).

Focal surfaces in differential geometry are also described with the help of line congruences. The definition of line congruence is first explained in visualization by Hagen et al. (1991). A line congruence is a set of lines given by two parameters in $3-$ dimensional space. Let $M\colon F(s,t)$ be a surface and its unit normal vector be denoted by $N(s,t)$. The line congruence is parameterized by

$$R(s,t,\lambda)=F(s,t)+\lambda\xi(s,t) \tag{3.1}$$

where $\xi(s,t)$ is the unit vectors on each points on M and $\lambda \in IR$. If $\xi(s,t)=N(s,t)$, then $R=R_N$ is called normal congruence, and also this representation is congruent to the focal surface.

 DOI: 10.1201/9781003322856-3

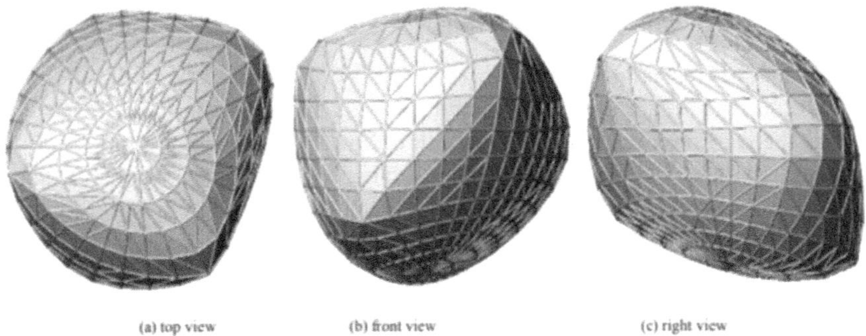

FIGURE 3.1 Some views of a discrete model.

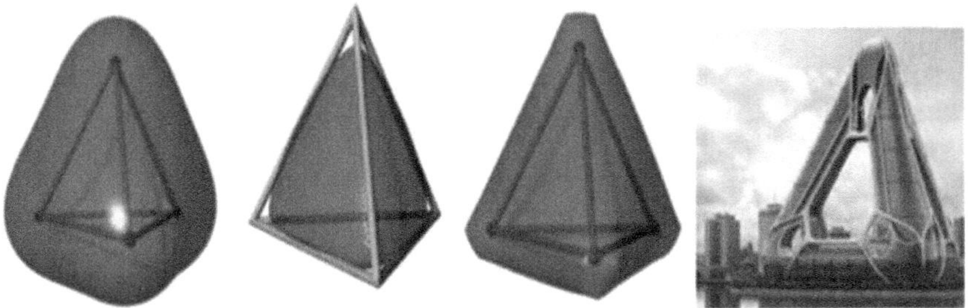

FIGURE 3.2 Generated surface and conceptual design of megastructure.

Thus, the focal surface R_N is defined by

$$F_1^*(s,t) = F(s,t) + \kappa_i^{-1}(s,t) N(s,t) \tag{3.2}$$

where κ_1 and κ_2 indicate the principle curvatures of M: $F(s,t)$[4]. Line congruences are known as a set of lines tangent to two surfaces. The two surfaces mentioned here are focal surfaces. Curvature centers of two principal directions correspond to focal points of the normal congruence. In literature, there exist many studies on focal curves and focal surfaces in Euclidean spaces [5–8].

A special surface named a canal surface has many properties in geometry. These types of surfaces are combinations of variable-radius moving spheres. They are used in producing images of internal organs, surface modeling, computer graphics, and Computer Aided Design (CAD). To describe these surfaces; $\gamma(s)$ is the curve representing the location of the centers of the spheres producing the canal surface, and $r(s)$ is known as the spheres' radius function. If the radius function is chosen as constant, then this corresponds to the tube surface.

Tube surfaces, which have been studied before and which are also the subject of this chapter, are structurally different, as they are a special case of canal surfaces [9]. In [10], the author focused on tube surfaces in $3-$ dimensional space. Last, in [11–14], the authors studied tube surfaces in $4-$ dimensional space, IE^4. Due to the geometric structure of tube surfaces, they have some applications for robots and their movements (see [15]).

Frame fields are important concepts for classifying surfaces in differential geometry. Despite the Frenet frame being the most well-known frame, other frame fields such as the Darboux frame are also preferred. Moreover, the Frenet frame is settled on a curve with acceleration vectors and its velocity. The Darboux frame is defined by the surface normal vector and the curves' velocity vector. There have been many studies about Darboux frames [4, 16, 17].

In this work, we investigate the focal surfaces of a tube surface that are constructed by the Frenet frame and Darboux frame. In Section 3.3, we yield the mean curvature and Gaussian curvature of the focal surfaces and find that the focal surface of a tube surface corresponds to a flat surface and these types of surfaces cannot be minimal. Furthermore, we present focal surface examples and give some results about s-parameter curves and t-parameter curves of related focal surfaces. In Section 3.4, we examine similar concepts with regard to the Darboux frame and get new results.

3.2 PRELIMINARIES

Let a unit speed curve in the Euclidean space $\mathbb{E}^3$ be denoted by $\gamma = \gamma(s)$ and let $\{T, N_1, N_2\}$ be the Frenet frame of γ. Then their derivatives (Frenet-Serret formula) are

$$\begin{bmatrix} T' \\ N_1' \\ N_2' \end{bmatrix} = \begin{bmatrix} 0 & \kappa & 0 \\ -\kappa & 0 & \tau \\ 0 & -\tau & 0 \end{bmatrix} \begin{bmatrix} T \\ N_1 \\ N_2 \end{bmatrix}$$

where κ, τ are the torsion and curvature of the curve γ, respectively.

Let a unit speed curve be indicated by $\gamma = \gamma(s) = I \rightarrow M$ on the surface M. A Darboux frame is defined along this curve and is given by $\{T, Y = T \times N, N\}$, where N is the unit normal vector of M, and T is the tangent vector of γ. Then, the formulas of the Darboux frame are

$$\begin{bmatrix} T' \\ Y' \\ N' \end{bmatrix} = \begin{bmatrix} 0 & k_g & k_n \\ -k_g & 0 & \tau_g \\ -k_n & -\tau_g & 0 \end{bmatrix} \begin{bmatrix} T \\ Y \\ N \end{bmatrix}$$

where k_n is the normal curvature, k_g is the geodesic curvature, and τ_g is the geodesic torsion of the curve γ.

The Darboux frame and Frenet frame are related with

$$\begin{bmatrix} T \\ Y \\ N \end{bmatrix} = \begin{bmatrix} 1 & 0 & 0 \\ 0 & \cos\theta & \sin\theta \\ 0 & -\sin\theta & \cos\theta \end{bmatrix} \begin{bmatrix} T \\ N_1 \\ N_2 \end{bmatrix}$$

Here, the angle between Y and N_1 is indicated by θ, and Darboux curvatures are expressed as $k_g = \kappa\cos\theta$, $k_n = \kappa\sin\theta$, and $\tau_g = \tau - \theta'$.

Let a regular surface in $\mathbb{E}^3$ be denoted by $M: F(s,t)$. Then, F_s and F_t span the tangent space of M at a point $p = F(s,t)$. The first fundamental form coefficients of M are given with the help of the Euclidean inner product:

$$e = \langle F_s, F_s \rangle, \quad f = \langle F_s, F_t \rangle, \quad g = \langle F_t, F_t \rangle \tag{3.3}$$

We set $W^2 = eg - f^2 \neq 0$.

The unit normal vector field of M is calculated by:

$$N = \frac{F_s \times F_t}{\| F_s \times F_t \|} \tag{3.4}$$

The second fundamental form coefficients are as follows:

$$l = \langle F_{ss}, N \rangle, \quad m = \langle F_{st}, N \rangle, \quad n = \langle F_{tt}, N \rangle \tag{3.5}$$

[18].

The following relation gives the shape operator matrix of M:

$$A_N = \begin{bmatrix} \dfrac{l}{e} & \dfrac{1}{W}\left(m - \dfrac{f}{e}l\right) \\ \dfrac{1}{W}\left(m - \dfrac{f}{e}l\right) & \dfrac{1}{W^2}\left(en - 2fm + \dfrac{f^2}{e}l\right) \end{bmatrix}$$

[19].

The mean curvature H and Gaussian curvature K are given by

$$H = \frac{en + gl - 2fm}{2\left(eg - f^2\right)} \tag{3.6}$$

and

$$K = \frac{\ln - m^2}{eg - f^2} \tag{3.7}$$

respectively [18, 20].

The unit speed curve $\gamma = I \to M$ on the surface M corresponds to an asymptotic curve if the normal curvature is zero in the direction γ'. In addition, the necessary and sufficient condition for γ to be an asymptotic curve is that $\gamma'' \perp N$. Furthermore, if the tangential component $(\gamma'')^T$ of the acceleration of γ is zero, then γ is a geodesic curve on M [18].

3.3 FOCAL SURFACE OF A TUBE SURFACE WITH FRENET FRAME

A tube surface with a Frenet frame was considered by Öztürk et al. in [9]. In this part, we deal with the tube surface with a Frenet frame and present the focal surface of this surface in IE^3.

Assume that a curve γ is parametrized by arclength as $\gamma(s) = (\gamma_1(s), \gamma_2(s), 0) \subset E^3$. Then the famous Frenet formulas become

$$\begin{aligned} \gamma' &= T \\ T' &= \kappa N_1 \\ N_1' &= -\kappa T \\ N_2' &= 0 \end{aligned}$$

The tube surface with regard to the Frenet frame has the parametrization:

$$M: F(s,t) = \gamma(s) + r(\cos t N_1(s) + \sin t N_2) \tag{3.8}$$

Here the radius of the spheres $r = const.$ F_s and F_t span the tangent space of M at a point $p = F(s,t)$ and are given by

$$\begin{aligned} F_s &= (1 - \kappa(s) r \cos t) T \\ F_t &= -r \sin t N_1 + r \cos t N_2 \end{aligned} \tag{3.9}$$

Then the coefficients of the first fundamental form are

$$e = (1 - \kappa(s) r \cos t)^2, \quad f = 0, \quad g = r^2 \tag{3.10}$$

where $W^2 = eg - f^2 = (1 - \kappa(s) r \cos t)^2 r^2$ [9].

Proposition 3.1 *[9] The tube surface $F(s,t)$ is regular if and only if $1 - \kappa(s) r \cos t \neq 0$.*

The second partial derivatives and the unit normal vector field of M are obtained as

$$\begin{aligned} F_{ss} &= -\kappa'(s) r \cos t T + \kappa(s)(1 - \kappa(s) r \cos t) N_1 \\ F_{st} &= \kappa(s) r \sin t T \\ F_{tt} &= -r \cos t N_1 - r \sin t N_2 \end{aligned} \tag{3.11}$$

and

$$N = -\cos t N_1 - \sin t N_2$$

respectively.

Then the second fundamental form coefficients are

$$l = -\kappa(s)(1 - \kappa(s) r \cos t) \cos t, \quad m = 0, \quad n = r \tag{3.12}$$

Therefore, from equations (3.10) and (3.12), the Gaussian and the mean curvature functions of M are obtained as

$$K = \frac{-\kappa(s) \cos t}{r(1 - \kappa(s) r \cos t)}$$

and

$$H = \frac{1 - 2\kappa(s) r \cos t}{2r\left(1 - \kappa(s) r \cos t\right)}$$

The following relation gives the shape operator matrix of the surface M:

$$A_N = \begin{bmatrix} \dfrac{-\kappa(s)\cos t}{1 - \kappa(s) r \cos t} & 0 \\ 0 & \dfrac{1}{r} \end{bmatrix} \tag{3.13}$$

[8].

Now, we can present the parametrization of the focal surface M^* of M by yielding the principle curvature functions of M.

By the use of (3.13), we obtain the principle curvature functions as

$$\kappa_1 = \frac{1}{r}, \quad \kappa_2 = \frac{-\kappa(s)\cos t}{1 - \kappa(s) r \cos t} \tag{3.14}$$

With the help of the definition of the focal surface of a given surface, and using (3.14), we get the focal surface $M*$ of M as

$$F^*(s,t) = \gamma(s) + \frac{1}{\kappa(s)\cos t}\left(\cos t N_1(s) + \sin t N_2\right) \tag{3.15}$$

The tangent vector fields of $M*$ are

$$\left(F^*\right)_s = \frac{-\kappa'(s)}{\kappa^2(s)}\left(N_1 + \tan t N_2\right) \tag{3.16}$$

and

$$\left(F^*\right)_t = \frac{1}{\kappa(s)\cos^2 t} N_2$$

Hence from (3.16), the first fundamental form coefficients are yielded as

$$e^* = \frac{\left(\kappa'(s)\right)^2}{\kappa^4(s)\cos^2 t}, \quad f^* = \frac{-\kappa'(s)\sin t}{\kappa^3(s)\cos^3 t}, \quad g^* = \frac{1}{\kappa^2(s)\cos^4 t} \tag{3.17}$$

where $\left(W^*\right)^2 = \dfrac{\left(\kappa'(s)\right)^2}{\kappa^6(s)\cos^4 t} \neq 0.$

Further, from (3.9), we find

$$N^* = -T \tag{3.18}$$

where $N*$ is the normal vector of M^*.

The second partial derivatives of $F^*(s,t)$ are

$$(F^*)_{ss} = \frac{\kappa'(s)}{\kappa(s)}T + \frac{-\kappa''(s)\kappa(s)+2(\kappa'(s))^2}{\kappa^3(s)}N_1 + \frac{-\kappa''(s)\kappa(s)+2(\kappa'(s))^2}{\kappa^3(s)}\tan tN_2$$

$$(F^*)_{st} = -\frac{\kappa'(s)}{\kappa^2(s)}\left(1+\tan^2 t\right)N_2 \tag{3.19}$$

$$(F^*)_{tt} = \frac{2\sin t}{\kappa(s)\cos^3 t}N_2$$

Therefore, from (3.18) and (3.19), we obtain the second fundamental form coefficients of the focal surface M^* as

$$l^* = -\frac{\kappa'(s)}{\kappa(s)}, \quad m^* = 0, \quad n^* = 0 \tag{3.20}$$

Example 3.1 *Let us handle the unit speed planar curve with the parametrization*

$$\gamma(s) = \left(\left(\frac{s}{\sqrt{2}}+1\right)\cos\left(\ln\left(\frac{s}{\sqrt{2}}+1\right)\right), \left(\frac{s}{\sqrt{2}}+1\right)\sin\left(\ln\left(\frac{s}{\sqrt{2}}+1\right)\right), 0\right)$$

The Frenet apparatus of this curve is determined by

$$T(s) = \gamma'(s) = \frac{1}{\sqrt{2}}\begin{pmatrix} \cos\left(\ln\left(\frac{s}{\sqrt{2}}+1\right)\right) - \sin\left(\ln\left(\frac{s}{\sqrt{2}}+1\right)\right), \\ \sin\left(\ln\left(\frac{s}{\sqrt{2}}+1\right)\right) + \cos\left(\ln\left(\frac{s}{\sqrt{2}}+1\right)\right), 0 \end{pmatrix}$$

$$N_1(s) = \frac{1}{\sqrt{2}}\begin{pmatrix} \sin\left(\ln\left(\frac{s}{\sqrt{2}}+1\right)\right) - \cos\left(\ln\left(\frac{s}{\sqrt{2}}+1\right)\right), \\ \cos\left(\ln\left(\frac{s}{\sqrt{2}}+1\right)\right) - \sin\left(\ln\left(\frac{s}{\sqrt{2}}+1\right)\right), 0 \end{pmatrix}$$

$$N_2(s) = (0, 0, 1)$$

$$\kappa(s) = \frac{1}{s+\sqrt{2}}, \quad \tau(s) = 0$$

Hence, the parametrization of a tubular surface around the curve $\gamma(s)$ can be written with the Frenet frame as

$$F(s,t)=\begin{pmatrix} \left(\frac{s}{\sqrt{2}}+1\right)\cos\left(\ln\left(\frac{s}{\sqrt{2}}+1\right)\right)-\frac{r}{\sqrt{2}}\cos t\left(\sin\left(\ln\left(\frac{s}{\sqrt{2}}+1\right)\right)+\cos\left(\ln\left(\frac{s}{\sqrt{2}}+1\right)\right)\right), \\ \left(\frac{s}{\sqrt{2}}+1\right)\sin\left(\ln\left(\frac{s}{\sqrt{2}}+1\right)\right)+\frac{r}{\sqrt{2}}\cos t\left(\cos\left(\ln\left(\frac{s}{\sqrt{2}}+1\right)\right)-\sin\left(\ln\left(\frac{s}{\sqrt{2}}+1\right)\right)\right), \\ r\sin t \end{pmatrix}$$

Also, the parametrization of the focal surface of this surface is yielded as

$$F^*(s,t)=\begin{pmatrix} \left(\frac{s^2+2\sqrt{2}s+1}{\sqrt{2}\left(s+\sqrt{2}\right)}\right)\cos\left(\ln\left(\frac{s}{\sqrt{2}}+1\right)\right)-\frac{1}{\sqrt{2}\left(s+\sqrt{2}\right)}\sin\left(\ln\left(\frac{s}{\sqrt{2}}+1\right)\right), \\ \left(\frac{s^2+2\sqrt{2}s+1}{\sqrt{2}\left(s+\sqrt{2}\right)}\right)\sin\left(\ln\left(\frac{s}{\sqrt{2}}+1\right)\right)+\frac{1}{\sqrt{2}\left(s+\sqrt{2}\right)}\cos\left(\ln\left(\frac{s}{\sqrt{2}}+1\right)\right), \\ \frac{1}{s+\sqrt{2}}\tan t \end{pmatrix}$$

Moreover, by substituting radius $r=\sqrt{2}$, the tube surface and its focal surface can be plotted in IE^3:

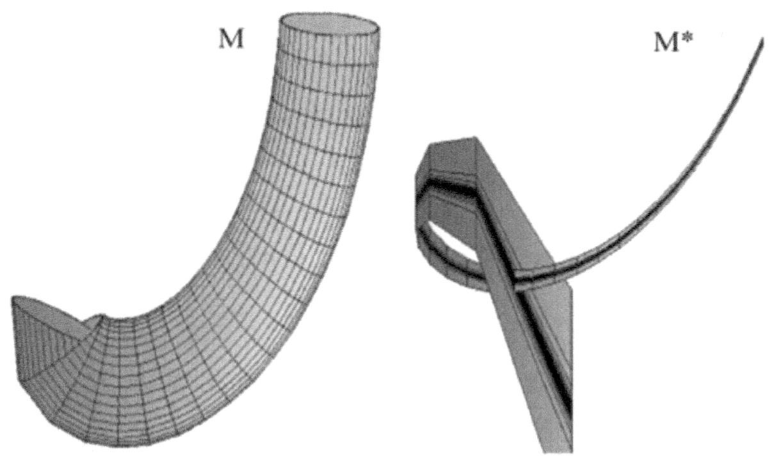

FIGURE 3.3 Tube surface M and the focal surface M^*,

Now, we can present the following results.

Theorem 3.1 *Let a tube surface M in $\mathbb{E}^3$ be parametrized by (3.8) and the focal surface M^* be given by (3.15). Then the Gaussian curvature of M^* is zero; hence the focal surface is congruent to a flat surface. In addition, the mean curvature of M^* is*

$$H^* = -\frac{\kappa^3(s)}{2\kappa'(s)} \tag{3.21}$$

Proof. Suppose that the focal surface M^* of the tube surface M is parametrized by (3.15) in $\mathbb{E}^3$. By the use of equations (3.6), (3.7), (3.17), and (3.20), we yield $K^* = 0$ and (3.21). This completes the proof.

Corollary 3.1 *There is no minimal focal surface M^* in $\mathbb{E}^3$.*

Proof. Let the focal surface M^* be a minimal surface. With the help of equation (3.21), the curvature function κ with respect to the Frenet frame is zero; this is a contradiction. Therefore, a focal surface M^* cannot be minimal.

Theorem 3.2 *Let a tube surface M in $\mathbb{E}^3$ be parametrized by (3.8) and the focal surface M^* be given by (3.15). Then the focal surface M^* has constant mean curvature if and only if the curvature function of $\kappa(s)$ satisfies*

$$\kappa(s) = \pm\frac{\sqrt{(s+c_1c)c}}{s+c_1c} \tag{3.22}$$

where c, c_1 are real constants.

Proof. Let the focal surface M^* be parametrized by (3.8) in $\mathbb{E}^3$. From equation (3.14), we obtain the differential equation

$$\kappa^3(s) + 2\kappa'(s)c = 0 \tag{3.23}$$

which has a non-trivial solution (3.22).

Now, we obtain the following results for parameter curves on the focal surface M^*.

Theorem 3.3 *Let a tube surface M in IE^3 be parametrized by (3.1) and the focal surface M^* be given by (3.8). Then*

1. *$s-$parameter curves of the focal surface M^* cannot be asymptotic curves.*
2. *$t-$parameter curves of the focal surface M^* are asymptotic curves.*

Proof. Let the focal surface M^* be parametrized by (3.8) in $\mathbb{E}^3$.

1. From the definition of an asymptotic curve, using (3.11) and (3.12), we get $\left\langle (F^*)_{ss}, N^* \right\rangle = \frac{\kappa'(s)}{\kappa(s)} = 0$ if and only if κ is constant. This is a contradiction because of the regularity. Hence, the $s-$parameter curves of the focal surface M^* cannot be asymptotic curves.

2. Similarly, by the use of the same equalities, we obtain $\left\langle \left(F^*\right)_{tt}, N^* \right\rangle = 0.$

   Therefore, we say the $t-$parameter curves of the focal surface M^* are asymptotic curves.

Theorem 3.4 *Let a tube surface M in $\mathbb{E}^3$ be parametrized by (3.1) and the focal surface M^* be given by (3.8). Then*

1. *$s-$parameter curves of the focal surface M^* are geodesic curves if and only if the equation*

$$\kappa(s) = -\frac{1}{c_1 s + c_2} \tag{3.24}$$

   holds for the curvature of γ, where c_1, c_2 are real constants.
2. *$t-$parameter curves of the focal surface M^* are geodesic curves if and only if $t = k\pi$, $k \in Z$.*

Proof. 1. From (3.18) and (3.19), we get $\left(F^*\right)_{ss} \wedge N^* = 0$ if and only if $-\kappa''\kappa + 2\left(\kappa'\right)^2 = 0$, which has a solution (3.24) for real constants c_1 and c_2.

2. Again using the same equations, we obtain $\left(F^*\right)_{tt} \wedge N^* = 0$ if and only if $\sin t = 0$, which completes the proof.

3.4 FOCAL SURFACE OF A TUBE SURFACE WITH DARBOUX FRAME

A tube surface according to a Darboux frame was handled by Yaylı and Doğan in [21]. In the present part, we deal with the tube surface with respect to the Darboux frame and present the focal surface of this type of surface in $\mathbb{E}^3$.

Assume that a curve γ on a surface S is parametrized by arclength as $\gamma(s) = \left(\gamma_1(s), \gamma_2(s), \gamma_3(s)\right) \subset IE^3$. The tubular surface with respect to the Darboux frame has the representation

$$M: F(s,t) = \gamma(s) + r\left(\cos tY(s) + \sin tU(s)\right) \tag{3.25}$$

Here the radius of the spheres; $r = const.$ and U is the unit normal of the surface S along the curve γ. Then F_s and F_t span the tangent space at a point $p = F(s,t)$:

$$\begin{aligned} F_s &= (1 - br)T - r\tau_g \sin tY + r\tau_g \cos tU \\ F_t &= -r\sin tY + r\cos tU \end{aligned} \tag{3.26}$$

where

$$b(s,t) = k_g(s)\cos t + k_n(s)\sin t \tag{3.27}$$

Then coefficients of the first fundamental form become

$$e=\left(1-br\right)^2+r^2\tau_g^2,\quad f=r^2\tau_g,\quad g=r^2 \tag{3.28}$$

where $W^2=eg-f^2=\left(1-br\right)^2r^2$ [18].

Proposition 3.2 *[18] The tube surface $F\left(s,t\right)$ is regular if and only if $b\neq\frac{1}{r}$.*

The second partial derivatives and the unit normal vector field of M are yielded as

$$\begin{aligned}F_{ss}&=(-b_s r-r\tau_g b_t)T+(k_g(1-br)-r\tau_g{}'\sin t-r\tau_g^2\cos t)Y\\&\qquad+(k_n(1-br)+r\tau_g{}'\cos t-r\tau_g^2\sin t)U\\F_{st}&=-b_t rT-r\tau_g\cos tY-r\tau_g\sin tU\\F_{tt}&=-r\cos tY-r\sin tU\end{aligned} \tag{3.29}$$

and

$$N=-\cos tY-\sin tU \tag{3.30}$$

Then the second fundamental form coefficients are

$$l=-(1-br)b+r\tau_g^2,\quad m=r\tau_g,\quad n=r \tag{3.31}$$

Therefore, from equations (3.28) and (3.31), the Gaussian and the mean curvature functions of M are obtained as

$$K=\frac{b}{r\left(br-1\right)} \tag{3.32}$$

and

$$H=\frac{1-2br}{2r(1-br)} \tag{3.33}$$

respectively [18].

Now, we focus on the representation of the focal surface M^* of M by obtaining the principal curvature functions of M.

Using (3.32), (3.33), and the equation $\kappa_i=H\pm\sqrt{H^2-K},(i=1,2)$, we get the principal curvature functions as

$$\kappa_1=\frac{1}{r},\quad \kappa_2=\frac{b}{br-1} \tag{3.34}$$

With the help of the definition of the focal surface of a given surface and using (4.10), we get the focal surface M^* as

$$F^*\left(s,t\right)=\gamma\left(s\right)+\frac{1}{b(s,t)}\left(\cos tY\left(s\right)+\sin tU(s)\right) \tag{3.35}$$

where $b(s,t)=k_g(s)\cos t+k_n(s)\sin t$.

The tangent vector fields of M^* are

$$\left(F^*\right)_s=\left(-\frac{b_s}{b^2}\cos t-\frac{1}{b}\tau_g\sin t\right)Y+\left(-\frac{b_s}{b^2}\sin t+\frac{1}{b}\tau_g\cos t\right)U \tag{3.36}$$

and

$$\left(F^*\right)_t=\left(-\frac{b_t}{b^2}\cos t-\frac{1}{b}\sin t\right)Y+\left(-\frac{b_t}{b^2}\sin t+\frac{1}{b}\cos t\right)U \tag{3.37}$$

Hence, from (3.36) and (3.37), the first fundamental form coefficients are yielded as follows:

$$e^*=\frac{b_s^2+b^2\tau_g^2}{b^4},\quad f^*=\frac{b_sb_t+b^2\tau_g}{b^4},\quad g^*=\frac{b_t^2+b^2}{b^4} \tag{3.38}$$

where $\left(W^*\right)^2=\frac{1}{b^6}(b_s-b_t\tau_g)^2$.

Proposition 3.3 *The focal surface* $M^*:F^*(s,t)$ *is regular if and only if* $(k_g'-k_n\tau_g)\cos t+(k_n'+k_g\tau_g)\sin t\neq 0$.

Proof. Substituting the partial derivatives of equation (3.27) in W^*, we get the result.

One can find the normal vector of the focal surface M^* as

$$N^*=T \tag{3.39}$$

The second partial derivatives of $F^*(s,t)$ are

$$\begin{aligned}\left(F^*\right)_{ss}&=\left(\frac{b_s-b_t\tau_g}{b}\right)T\\&+\left(\frac{-b_{ss}b^2\cos t+2bb_s^2\cos t+2b_sb^2\tau_g\sin t-b^3\tau_g^2\cos t-b^3\tau_g'\sin t}{b^4}\right)Y\\&+\left(\frac{-b_{ss}b^2\sin t+2bb_s^2\sin t-2b_sb^2\tau_g\cos t-b^3\tau_g^2\sin t+b^3\tau_g'\cos t}{b^4}\right)U\end{aligned}$$

$$\begin{aligned}\left(F^*\right)_{st}&=\left(\frac{-b_{st}b^2\cos t+2bb_sb_t\cos t+b_sb^2\sin t+b_tb^2\tau_g\sin t-b^3\tau_g\cos t}{b^4}\right)Y\\&+\left(\frac{-b_{st}b^2\sin t+2bb_sb_t\sin t-b_sb^2\cos t-b_tb^2\tau_g\cos t-b^3\tau_g\sin t}{b^4}\right)U\end{aligned}$$

$$\left(F^*\right)_{tt} = \left(\frac{-b_{tt}b^2\cos t + 2bb_t^2\cos t + 2b_t b^2\sin t - b^3\cos t}{b^4}\right)Y + \left(\frac{-b_{tt}b^2\sin t + 2bb_t^2\sin t - 2b_t b^2\cos t - b^3\sin t}{b^4}\right)U \tag{3.40}$$

Therefore, from (3.38) and (3.40), we obtain the second fundamental form coefficients of the focal surface M^* as

$$l^* = \frac{b_s - b_t\tau_g}{b}, \quad m^* = 0, \quad n^* = 0 \tag{3.41}$$

Example 3.2 *Let us consider the unit speed curve with the parametrization.*

$$\gamma(s) = \left(\cos\frac{s}{\sqrt{2}}, \sin\frac{s}{\sqrt{2}}, \frac{s}{\sqrt{2}}\right)$$

The Darboux frame vectors and the curvatures k_g, k_n of this curve are obtained by

$$T(s) = \gamma'(s) = \frac{1}{\sqrt{2}}\left(-\sin\frac{s}{\sqrt{2}}, \cos\frac{s}{\sqrt{2}}, 1\right)$$

$$Y(s) = \frac{1}{\sqrt{2}}\left(-\cos\frac{s}{\sqrt{2}} + \frac{1}{\sqrt{2}}\sin\frac{s}{\sqrt{2}}, -\sin\frac{s}{\sqrt{2}} - \frac{1}{\sqrt{2}}\cos\frac{s}{\sqrt{2}}, \frac{1}{\sqrt{2}}\right)$$

$$U(s) = \frac{1}{\sqrt{2}}\left(\cos\frac{s}{\sqrt{2}} + \frac{1}{\sqrt{2}}\sin\frac{s}{\sqrt{2}}, \sin\frac{s}{\sqrt{2}} - \frac{1}{\sqrt{2}}\cos\frac{s}{\sqrt{2}}, \frac{1}{\sqrt{2}}\right)$$

$$k_g(s) = \frac{1}{2\sqrt{2}}, \quad k_n(s) = \frac{1}{2\sqrt{2}}$$

Hence, the tube surface's parametrization (around the curve $\gamma(s)$) can be given according to the frame as

$$F(s,t) = \begin{pmatrix} \cos\frac{s}{\sqrt{2}}(1 - \cos t + \sin t) + \frac{1}{\sqrt{2}}\sin\frac{s}{\sqrt{2}}(\cos t + \sin t), \\ \sin\frac{s}{\sqrt{2}}(1 - \cos t + \sin t) - \frac{1}{\sqrt{2}}\cos\frac{s}{\sqrt{2}}(\cos t + \sin t), \\ \frac{s + \cos t + \sin t}{\sqrt{2}} \end{pmatrix}$$

In addition, the parametrization of $M*$ is yielded as

$$F^*(s,t) = \begin{pmatrix} \cos\frac{s}{\sqrt{2}}\left(\frac{3\sin t - \cos t}{\cos t + \sin t}\right) + \sqrt{2}\sin\frac{s}{\sqrt{2}}, \\ \sin\frac{s}{\sqrt{2}}\left(\frac{3\sin t - \cos t}{\cos t + \sin t}\right) - \sqrt{2}\cos\frac{s}{\sqrt{2}}, \frac{s+2}{\sqrt{2}} \end{pmatrix}$$

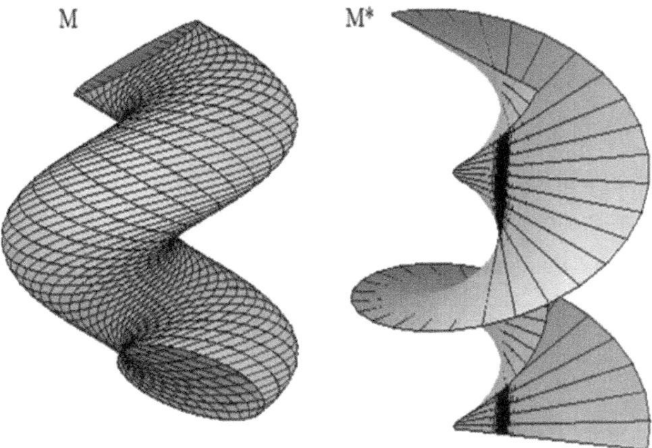

FIGURE 3.4 Tubular surface M and focal surface M^*.

Moreover, by substituting radius $r = \sqrt{2}$, the tube surface and its focal surface can be plotted in IE^3:

Now, we can present the following ramifications.

Theorem 3.5 *Let a tube surface M with respect to a Darboux frame in $\mathbb{E}^3$ be parametrized by (3.25) and the focal surface M^* be given by (3.35). Then the Gaussian curvature of M^* is zero, so the focal surface corresponds to a flat surface.*

Proof. Suppose that the focal surface M^* of the tube surface M is parametrized by (3.35) in IE^3. By the use of equations (3.38), (3.41), and (3.7), we yield $K^* = 0$. This completes the proof.

Theorem 3.6 *Let a tube surface M with respect to a Darboux frame in $\mathbb{E}^3$ be parametrized by (3.25) and the focal surface M^* be given by (3.35). Then the mean curvature of M^* is*

$$H^* = \frac{(b_t^2 + b^2)b}{2(b_s - b_t \tau_g)} \tag{3.42}$$

Proof. Suppose that the focal surface M^* of the tube surface M is parametrized by (3.35) in IE^3. By the use of equations (3.6), (3.38), and (3.41), we get the result.

Theorem 3.7 *There is no minimal focal surface M^* given by (3.35) in $\mathbb{E}^3$.*

Proof. Let the focal surface M^* be a minimal surface. With the help of equation (3.42), we yield $H^* = 0 \Leftrightarrow b = 0$. This a contradiction. Therefore, the focal surface M^* cannot be minimal.

Theorem 3.8 *Let a tube surface M with a Darboux frame in $\mathbb{E}^3$ be parametrized by (3.25) and the focal surface M^* be given by (3.35). Then*

1. *$s -$ parameter curves of the focal surface M^* cannot be asymptotic curves.*
2. *$t -$ parameter curves of the focal surface M^* are asymptotic curves.*

Proof. Let the focal surface M^* be parametrized by (3.35) in IE^3.

1. Using equations (3.39) and (3.40) and the definition of an asymptotic curve, we get $\left\langle \left(F^*\right)_{ss}, N^* \right\rangle = 0$ if and only if $b_s - b_t \tau_g = 0$. This is a contradiction because of the regularity. Hence, the $s-$parameter curves of the focal surface M^* cannot be asymptotic curves.
2. Similarly, by the use of the same equalities, we obtain $\left\langle \left(F^*\right)_{tt}, N^* \right\rangle = 0$. Therefore, we say the $t-$parameter curves of the focal surface M^* are asymptotic curves.

Theorem 3.9 *Let a tube surface M with respect to a Darboux frame in $\mathbb{E}^3$ be parametrized by (3.25) and the focal surface M^* be given by (3.35). Then,*

1. *$s-$parameter curves of the focal surface M^* are geodesic curves if and only if*
$$\tau_g(s) = \frac{4c_1}{c_1^2(s+c_2)^2 + 16}$$
where c_1, c_2 are real constants.
2. *$t-$parameter curves of the focal surface M^* cannot be geodesic curves.*

Proof. 1. From (3.39) and the second partial derivative of $F^*(s,t)$, $\left(F^*\right)_{ss} \wedge N^* = 0$ if and only if

$$b(-b_{ss}b + 2b_s^2 - b^2\tau_g^2) = 0 \tag{3.43}$$

$$b^2(2b_s\tau_g - b\tau_g{}') = 0 \tag{3.44}$$

In equality (3.44), since $b \neq 0$, we know $2\frac{b_s}{b} = \frac{\tau_g{}'}{\tau_g}$. Therefore, we get

$$b^2 = \tau_g(s)h(t)$$

where $h(t)$ is a smooth function. According to the relation between this equation and (3.43), we have the differential equation

$$2\tau_g{}''\tau_g - 3(\tau_g{}')^2 + 4\tau_g^4 = 0$$

which has the solution (3.42). Thus, we obtain the desired result.

2. Similarly, $\left(F^*\right)_{tt} \wedge N^* = 0$ if and only if

$$b(b_{tt}b - 2b_t^2 + b^2) = 0$$
$$b_t = 0$$

FIGURE 3.5 Some architectural examples around the world.

Using these relations, we get $b = 0$. This is a contradiction because of the parametrization of the focal surface. Hence, we get the result, and this completes the proof.

3.5 CONCLUSION

Rotational surfaces (especially canal surfaces) have attracted attention in terms of their applications. The tubular surface, which is a special channel surface, has many applications in robots and their movements. In addition to being used in producing images of internal organs, computer graphics, and CG/CAD, canal surfaces and tubular surfaces are also used in surface modeling.

Surface modeling is the most important stage in transferring surfaces to architecture. In the past, many models have been used in architecture based on surface examples in geometry. Some of the architectural examples around the world are the Colosseum in Rome, Italy; an elliptical-based house, Buenos Aires, Argentina; and a cathedral with hyperbolic arches, Brazil [15]. Figure 3.5 shows some views of these architectural structures:

Focal surfaces are generally considered theoretically by geometers. In [2, 6–8], the authors study focal surfaces in three-dimensional and four-dimensional spaces.

In this work, we handle a tubular surface in 3-dimensional Euclidean space and define the focal surface of this surface with respect to the Frenet frame and Darboux frame. After obtaining the parameterization of the surface, we calculate Gaussian and mean curvatures. Then, we get the conditions for the focal surface of a tube surface to become flat and minimal. We characterize $s-$parameter curves and $t-$parameter curves of these types of surfaces. We present some examples of these types of surfaces and plot them by the use of Maple. (Mathematical software for visualization and solving problems.)

In this chapter, we deal with the focal surface of a tube surface more theoretically. Based on the surface examples and figures given in this chapter, surface modeling can be obtained and applied in different fields, such as architecture.

REFERENCES

[1] H. Hagen, H. Pottmann, A. Divivier, Visualization Functions on a Surface, *Journal of Visualization and Animation*, 2 (1991), 52–58.

[2] L. Petrusevski, M. Petrovic, M. Devetakovic, J. Ivanovic, Modelling of Focal-Directional Surfaces for Application in Architecture, *FME Transactions*, 45 (2017), 294–300.

[3] M. Petrovic, *Generating the Focal Directorial Geometric Forms as a Designing Pattern of the Architectural-Urban Space (in Serbian)*, Doctoral Dissertation, University of Belgrade, Faculty of Architecture, 2016.

[4] M. Düldül, B. Uyar Düldül, N. Kuruoğlu, E. Özdamar, Extension of the Darboux Frame into Eucliden 4-Space and Its Invariants, *Turkish Journal of Mathematics*, 41 (2017), 1628–1639.

[5] H. Hagen, S. Hahmann, Generalized Focal Surfaces, *A New Method for Surface Interrogation, Proceedings Visualization'92, Boston* (1992), 70–76.

[6] B. Özdemir, *A Characterization of Focal Curves and Focal Surfaces in IE^n*, PhD Thesis, Uludag University, Bursa, Turkey, 2008.

[7] B. Özdemir, K. Arslan, On Generalized Focal Surfaces in IE^3, *Review Bulletin of the Calcutta Mathematical Society*, 16 (2008), 23–32.

[8] G. Öztürk, K. Arslan, On Focal Curves in Euclidean n-Space IR^n, *Novi Sad Journal of Mathematics*, 48 (2016), 35–44.

[9] G. Öztürk, B. Bulca, B.K. Bayram, K. Arslan, On Canal Surfaces in IE^3, *Selçuk Journal of Applied Mathematics*, 11 (2010), 103–108.

[10] A.H. Sorour, Weingarten Tube-Like Surfaces in Euclidean 3-Space, *Studia Universitatis Babeş-Bolyai Mathematica*, 61 (2016), 239–250.

[11] B. Bulca, K. Arslan, B. Bayram, G. Öztürk, Canal Surfaces in 4-Dimensional Euclidean Space, *An International of Optimization and Control: Theories & Applications*, 7(1) (2017), 83–89.

[12] I. Kişi, G. Öztürk, A New Approach to Canal Surface with Parallel Transport Frame, *International Journal of Geometric Methods in Modern Physics*, 14 (2017), 1–16.

[13] I. Kişi, G. Öztürk, A New Type of Tubular Surface Having Pointwise 1-Type Gaussmap in Euclidean 4-Space IE^4, *Journal of the Korean Mathematica Society*, 55 (2018), 923–938.

[14] I. Kişi, G. Öztürk, K. Arslan, A New Type of Canal Surface in Euclidean Space IE^4, *Sakarya University Journal of Science*, 23 (2019), 801–809.

[15] T. Maekawa, M.N. Patrikalakis, T. Sakkalis, G. Yu, Analysis and Applications of Pipe Surfaces, *Computer Aided Geometric Design*, 15 (1998), 437–458.

[16] Y. Kemer, E. Ata, Frenet-Serret and Darboux Frames of a Image Curve on a Screw Surface, *Sinop Üniversitesi Fen Bilimleri Dergisi*, 2 (2017), 48–58.

[17] I. Kişi, M. Tosun, Spinor Darboux Equations of Curves in Euclidean 3-Space, *MathematicaMoravica*, 19 (2015), 87–93.

[18] P.M. Do Carmo, *Differential Geometry of Curves and Surfaces*, Englewood Cliffs, NJ: Prentice-Hall, 1976.

[19] B. Bulca, *A Characterization of Surfaces in IE^3*, Ph.D Thesis, Uludag University, Bursa, 2012.

[20] E. İyigün, K. Arslan, G. Öztürk, A Characterization of Chen surfaces in IE^4, *Journal of the American Mathematical Society*, 31 (2008), 209–215.

[21] F. Doğan, Y. Yaylı, Tubes with Darboux Frame, *International Journal of Contemporary Mathematical Sciences*, 7 (2012), 751–758.

4 Approximation of Functions in Certain Lipschitz Classes by (M, λ_n) $(E, 1)$ Means of Fourier Series and Conjugate Series of Fourier Series

Shilpa Das and Hemen Dutta

4.1 INTRODUCTION

Summability methods are applicable as a tool for filtering of signals and stabilizing systems. Signals use to appear in the form of infinite series, Fourier series, wavelets, and so on. Summability methods can be applied to find the degree and error of approximation of such signals. This chapter aims to define first a product summability method by superimposing the (M, λ_n) method (also known as the Natarajan method) on the $(E, 1)$ method (also known as the Euler method) to use subsequently for obtaining the degree of approximation of a Fourier series of functions (signals) belonging to generalized Lipschitz classes $\text{Lip}(\alpha, r)$ and $\text{Lip}(\xi(t), r)$. $r \geq 1$. The method is further applied to obtain the degree of approximation of a conjugate series of Fourier series of functions belonging to the same Lipschitz classes.

Some findings on the degree of approximation of functions belonging to the Lipschitz classes and generalized Lipschitz classes by Cesàro, Nörlund, and the generalized Nörlund summability methods can be obtained from [1–13]. Also, some earlier findings on the error approximations through trigonometric Fourier approximation of the conjugate of a function in $\text{Lip}(\alpha, r)$ and $\text{Lip}(\xi(t), r)$ classes by product summability means $(C,1)(E,1)$, respectively, can be obtained from [14, 15]. The Fourier series and its conjugate series by $(C,1)(E,1)$ product means were studied in [16]. The work of [17] is concerned with the degree of approximation of function in the $\text{Lip}(\xi(t), r)$ class by $(C,1)(E,q)$ product summability means. The degree of approximation of functions in the $\text{Lip}(\alpha, r)$ and $\text{Lip}(\xi(t), r)$ classes by $(N, p_n)(E,1)$ product summability means of the Fourier series were carried out in [18]. The degree of approximation of functions (signals) in certain Lipschitz classes by the Zweier-Euler product summability method of the Fourier series was obtained in [19]. Results

DOI: 10.1201/9781003322856-4

on the degree of approximation of the functions in the classes $\mathrm{Lip}(\alpha,r)$ and Lip $(\xi(t),r)$ by the $(N,p_n)(E,1)$ means of conjugate series of the Fourier series were obtained in [20].

Corresponding to each continuous function on a closed interval and $\delta > 0$, there exists a polynomial which goes completely inside the band of width 2δ surrounding the graph of the function. A discontinuous function can never be approximated arbitrarily well by a polynomial. The approximation of a function using the Fourier series is called trigonometric approximation. Let $f(x)$ be a 2π-periodic function and integrable in the sense of Lebesgue over the interval $[0,2\pi]$. The trigonometric Fourier series associated with $f(x)$ is given by

$$f(x)=\frac{a_0}{2}+\sum_{n=1}^{\infty}(a_n\cos nx+b_n\sin nx)\equiv\frac{a_0}{2}+\sum_{n=1}^{\infty}A_n(x),\ n\in\mathbb{N} \tag{4.1}$$

with the nth partial sum $s_n(f,x)$.

The following one is the conjugate series of the previous Fourier series

$$\sum_{n=1}^{\infty}(b_n\cos nx-a_n\sin nx)\equiv\sum_{n=1}^{\infty}B_n(x) \tag{4.2}$$

A function $f\in\mathrm{Lip}\,\alpha$ if

$$f(x+t)-f(x)=O\left(|t|^{\alpha}\right),\ 0<\alpha\leq 1,\ t>0$$

$f\in\mathrm{Lip}(\alpha,r)$, for $0\leq x\leq 2\pi$, if

$$\left(\int_0^{2\pi}|f(x+t)-f(x)|^r\,dx\right)^{\frac{1}{r}}=O\left(|t|^{\alpha}\right),\ 0<\alpha\leq 1,\ t>0,\ r\geq 1$$

[8]

Let $\xi(t)$ be a positive increasing function and $r\geq 1$ be an integer. Then $f\in$ Lip $(\xi(t),r)$ if

$$\left(\int_0^{2\pi}|f(x+t)-f(x)|^r\,dx\right)^{\frac{1}{r}}=O(\xi(t))$$

We have $\|f\|_r=\left[\int_0^{2\pi}|f(x)|^r\,dx\right]^{\frac{1}{r}}$, $1\leq r<\infty$ (L_r-norm), and the degree of approximation $E_n(f)$ of $f\in L_r$ by trigonometric polynomial t_n of order n is

$$E_n(f)=\min_n\|f-t_n\|_r$$

[21]

Consider the series $\sum_{n=0}^{\infty} u_n$ with $s_n = \sum_{k=0}^{n} u_k$ (n^{th} partial sum). Also, consider $\{\lambda_n\}$, a sequence of non-negative numbers, real or complex, with $\sum_n |\lambda_n| < \infty$.

The transformation $t_n^M = \sum_{k=0}^{n} \lambda_{n-k} s_k = \sum_{k=0}^{n} \lambda_k s_{n-k}$ defines the sequence $\{t_n^M\}$ of Natarajan means of the sequence $\{s_n\}$ generated by the sequence of coefficients $\{\lambda_n\}$. If $t_n^M \to s$ as $n \to \infty$, we say that $\sum_{n=0}^{\infty} u_n$ is summable by Natarajan means or (M, λ_n) summable to the sum s.

The necessary and sufficient condition for regularity of the (M, λ_n) method is

$$\sum_n \lambda_n = 1$$

[22]

Let $E_n^{(1)} = \frac{1}{2^n} \sum_{k=0}^{n} \binom{n}{k} s_k$. If $E_n^{(1)} \to s$ as $n \to \infty$, then $\sum_{n=0}^{\infty} u_n$ is said to be summable to s by Euler method $(E, 1)$, which is a regular method [23].

The $(M, \lambda_n)(E, 1)$ method is obtained by superimposing the (M, λ_n) method on the $(E, 1)$ method.

The (M, λ_n) transform of the $(E, 1)$ transform defines the $(M, \lambda_n)(E, 1)$ transform t_n^{ME} of the nth partial sum s_n of the series $\sum_{n=0}^{\infty} u_n$ by

$$\begin{aligned} t_n^{ME} &= \sum_{k=0}^{n} \lambda_{n-k} E_k^{(1)} \\ &= \sum_{k=0}^{n} \lambda_{n-k} \frac{1}{2^k} \sum_{v=0}^{k} \binom{k}{v} s_v \\ &= \sum_{k=0}^{n} \lambda_k \frac{1}{2^{n-k}} \sum_{v=0}^{n-k} \binom{n-k}{v} s_v \end{aligned}$$

If $t_n^{ME} \to s$ as $n \to \infty$, then the series $\sum_{n=0}^{\infty} u_n$ is said to be $(M, \lambda_n)(E, 1)$ summable to the sum s.

The formula

$$\sum_{k=m}^{n} u_k v_k = \sum_{k=m}^{n-1} U_k \left(v_k - v_{k+1}\right) - U_{m-1} v_m + U_n v_n \tag{4.3}$$

where $0 \le m \le n$, $U_k = u_0 + u_1 + u_2 + \ldots + u_k$, if $k \ge 0$ and $U_{-1} = 0$, which can be checked, is known as Abel's transformation.

If $v_m, v_{m+1}, \ldots, v_n$ are non-negative and non-increasing, then the left-hand side of (4.3) does not exceed $2\, v_m \max_{m-1 \le k \le n} |U_K|$.

In fact,

$$\left|\sum_{k=m}^{n} u_k v_k\right| \le \max|U_k| \left\{\sum_{k=m}^{n-1} (v_k - v_{k+1}) + v_m + v_n\right\}$$

$$= 2\, v_m \max|U_k|$$

We use the following notations throughout this chapter

$$\varphi(x,t) = f(x+t) + f(x-t) - 2 f(x)$$

$$\varphi\sim(x,t) = f(x+t) - f(x-t)$$

$$\Delta\lambda_k = \lambda_k - \lambda_{k+1},\ k \ge 0$$

$$(ME)_n(t) = \frac{1}{2\pi}\sum_{k=0}^{n} \lambda_k \frac{cos^{n-k}\left(\frac{t}{2}\right)\sin(n-k+1)\left(\frac{t}{2}\right)}{\sin\left(\frac{t}{2}\right)}$$

and

$$(ME)_n^{\sim}(t) = \frac{1}{2\pi}\sum_{k=0}^{n} \lambda_k \frac{cos^{n-k}\left(\frac{t}{2}\right)\cos\left(n-k+\frac{1}{2}\right)(t)}{\sin\left(\frac{t}{2}\right)}$$

4.2 APPROXIMATION RESULTS CONCERNING FOURIER SERIES OF FUNCTIONS

In this section, we first give some lemmas that will be required to establish the main results of this section.

Lemma 4.1. *For* $0 < t \le \frac{\pi}{n+1}$, $(ME)_n(t) = O\left((n+1)^2\right)$.

Proof. For $0 < t \le \frac{\pi}{n+1}$, $\sin\left(\frac{t}{2}\right) \ge \frac{t}{\pi}$ and $\sin nt \le nt$, we have

$$\left|(ME)_n(t)\right| = \left|\frac{1}{2\pi}\sum_{k=0}^{n} \lambda_k \frac{cos^{n-k}\left(\frac{t}{2}\right)\sin(n-k+1)\left(\frac{t}{2}\right)}{\sin\left(\frac{t}{2}\right)}\right|$$

$$\leq \frac{1}{2\pi}\sum_{k=0}^{n} |\lambda_k| \frac{(n-k+1)\left(\frac{t}{2}\right)}{\frac{t}{\pi}}$$

$$= \frac{1}{4}\sum_{k=0}^{n} |\lambda_k|(n-k+1)$$

$$= O\left((n+1)^2\right)$$

Lemma 4.2. *For* $\frac{\pi}{n+1} < t \leq \pi$, $(ME)_n(t) = O\left(\frac{1}{t^2(n+1)}\right)$ *under the conditions* $(n+1)\lambda_n = O(1)$ *and* $\sum_{k=0}^{n-1}|\Delta\lambda_k| = O\left(\frac{1}{n+1}\right)$.

Proof. Suppose the conditions $(n+1)\lambda_n = O(1)$ and $\sum_{k=0}^{n-1}|\Delta\lambda_k| = O\left(\frac{1}{n+1}\right)$ hold.

For $\frac{\pi}{n+1} < t \leq \pi$, $\sin\left(\frac{t}{2}\right) \geq \frac{t}{\pi}$, $\left|cos^{n-k}\left(\frac{t}{2}\right)\right| \leq 1$, $0 < k < n$ and using Abel's lemma, we have

$$\left|(ME)_n(t)\right| - \left|\frac{1}{2\pi}\sum_{k=0}^{n}\lambda_k \frac{cos^{n-k}\left(\frac{t}{2}\right)\sin(n-k+1)\left(\frac{t}{2}\right)}{\sin\left(\frac{t}{2}\right)}\right|$$

$$\leq \frac{1}{2t}\sum_{k=0}^{n}\left|\lambda_k \sin(n-k+1)\left(\frac{t}{2}\right)\right|$$

$$\leq \frac{1}{2t}\left[\sum_{k=0}^{n-1}\left|\lambda_k - \lambda_{k+1}\right|\sum_{j=0}^{k}\sin(n-j+1)\left(\frac{t}{2}\right) + \lambda_n\sum_{k=0}^{n}\sin(n-k+1)\left(\frac{t}{2}\right)\right]$$

$$\leq \frac{1}{2t}\left[\sum_{k=0}^{n-1}|\lambda_k| + \lambda_0 + \lambda_n\right]\max_{0\leq k\leq n}\left|\sum_{j=0}^{k}\sin(n-j+1)\left(\frac{t}{2}\right)\right|$$

$$= O\left(\frac{1}{t^2}\right)\left[\sum_{k=0}^{n-1}|\Delta\lambda_k| + \lambda_0 + \lambda_n\right]$$

$$= O\left(\frac{1}{t^2}\right)\left(\frac{1}{n+1} + \frac{1}{n+1}\right)$$

$$= O\left(\frac{1}{t^2(n+1)}\right)$$

Lemma 4.3. [24] *Let* $f \in \text{Lip}(\alpha, r)$, $0 < \alpha \le 1$, $r \ge 1$, *then*

$$\left[\int_0^{2\pi} |\varphi(x,t)|^r \, dx\right]^{\frac{1}{r}} = O\left(|t|^\alpha\right)$$

Lemma 4.4. [19] *Let* $f \in \text{Lip}(\xi(t), r)$, $r \ge 1$, *then*

$$\left[\int_0^{2\pi} |\varphi(x,t)|^r \, dx\right]^{\frac{1}{r}} = O(\xi(t))$$

Theorem 4.1. Let (M, λ_n) be a regular method defined by a positive sequence $\{\lambda_n\}$ such that

$$(n+1)\lambda_n = O(1) \text{ and } \sum_{k=0}^{n-1} |\Delta\lambda_k| = O\left(\frac{1}{n+1}\right) \tag{4.4}$$

Let $f : [0, 2\pi] \to \mathbb{R}$ *be* 2π *-periodic, integrable in the sense of Lebesgue, and belonging to the class* $\text{Lip}(\alpha, r)$, $r \ge 1$, *then the degree of approximation of f by* $(M, \lambda_n)(E, 1)$ *transform*

$$t_n^{ME} = \sum_{k=0}^{n} \lambda_k \frac{1}{2^{n-k}} \sum_{v=0}^{n-k} \binom{n-k}{v} s_v$$

of its Fourier series (4.1) satisfies

$$\| t_n^{ME} - f \|_{L^r_{(\alpha)}} = O\left(\frac{1}{(n+1)^{\alpha-1}}\right) + \begin{cases} O\left(\dfrac{1}{(n+1)^\alpha}\right), & 0 < \alpha < 1 \\ O\left(\dfrac{\log(n+1)}{(n+1)}\right), & \alpha = 1. \end{cases}, \text{for } n = 0, 1, 2, 3, \ldots$$

Proof. By Titchmarsh [25] and using the Riemann-Lebesgue theorem, the nth partial sum $s_n(f, x)$ of the Fourier series (4.1) of $f(x)$ is given by

$$s_n(f,x) - f(x) = \frac{1}{2\pi} \int_0^\pi \varphi(x,t) \frac{\sin\left(n + \frac{1}{2}\right)t}{\sin\left(\frac{t}{2}\right)} dt$$

By $E_n^{(1)}(x)$, we denote $(E,1)$ means of $s_n(f,x)$, and we get

$$\frac{1}{2^n}\sum_{k=0}^{n}\binom{n}{k}\left(s_k(f,x)-f(x)\right)=\frac{1}{2^{n+1}\pi}\int_o^{\pi}\frac{\varphi(x,t)}{\sin\left(\frac{t}{2}\right)}\sum_{k=0}^{n}\binom{n}{k}\sin\left(k+\frac{1}{2}\right)tdt$$

$$E_n^{(1)}(x)-f(x)=\frac{1}{2^{n+1}\pi}\int_0^{\pi}\frac{\varphi(x,t)}{\sin\left(\frac{t}{2}\right)}\operatorname{Im}\left\{\sum_{k=0}^{n}\binom{n}{k}e^{i\left(k+\frac{1}{2}\right)t}\right\}dt$$

$$=\frac{1}{2^{n+1}\pi}\int_0^{\pi}\frac{\varphi(x,t)}{\sin\left(\frac{t}{2}\right)}\operatorname{Im}\left\{e^{\frac{it}{2}}\sum_{k=0}^{n}\binom{n}{k}e^{ikt}\right\}dt$$

$$=\frac{1}{2^{n+1}\pi}\int_0^{\pi}\frac{\varphi(x,t)}{\sin\left(\frac{t}{2}\right)}\operatorname{Im}\left\{e^{it/2}\left(1+e^{it}\right)^n\right\}dt$$

$$=\frac{1}{2^{n+1}\pi}\int_0^{\pi}\frac{\varphi(x,t)}{\sin\left(\frac{t}{2}\right)}\operatorname{Im}\left\{2^n cos^n\left(\frac{t}{2}\right)e^{i(n+1)t/2}\right\}dt$$

$$=\frac{1}{2\pi}\int_0^{\pi}\varphi(x,t)\frac{cos^n\left(\frac{t}{2}\right)\sin(n+1)\left(\frac{t}{2}\right)}{\sin\left(\frac{t}{2}\right)}dt$$

By $t_n^{ME}(x)$, we denote (M,λ_n) means of $E_n^{(1)}(x)$, and we obtain

$$\sum_{k=0}^{n}\lambda_k\left\{E_{n-k}^{(1)}(x)-f(x)\right\}=\sum_{k=0}^{n}\lambda_k\left\{\frac{1}{2\pi}\int_0^{\pi}\varphi(x,t)\frac{cos^{n-k}\left(\frac{t}{2}\right)\sin(n-k+1)\left(\frac{t}{2}\right)}{\sin\left(\frac{t}{2}\right)}dt\right\}$$

$$t_n^{ME}(x)-f(x)=\int_0^{\pi}\varphi(x,t)(ME)_n(t)dt$$

We have, by using Lemma 4.3 [26] (generalized Minkowski inequality)

$$\| t_n^{ME}-f\|_{L^r_{(\alpha)}}=\left[\int_0^{2\pi}\left|t_n^{ME}(x)-f(x)\right|^r dx\right]^{\frac{1}{r}}$$

$$=\left[\int_0^{2\pi}\left|\int_0^{\pi}\varphi(x,t)(ME)_n(t)dt\right|^r dx\right]^{\frac{1}{r}}$$

$$\leq \int_0^{\pi}\left\{\int_0^{2\pi} |\varphi(x,t)|^r\, dx\right\}^{\frac{1}{r}} |(ME)_n(t)|\, dt$$

$$= \int_0^{\pi} O\left(t^{\alpha}\right) |(ME)_n(t)|\, dt$$

$$= O\left(\int_0^{\frac{\pi}{n+1}} (t^{\alpha})(ME)_n(t)\, dt\right) + O\left(\int_{\frac{\pi}{n+1}}^{\pi} (t^{\alpha})(ME)_n(t)\, dt\right)$$

$$= I_1 + I_2, \text{ say} \tag{4.5}$$

Applying Lemma 4.1, we have

$$I_1 = O\left(\int_0^{\frac{\pi}{n+1}} (t^{\alpha})(n+1)^2\, dt\right)$$

$$= O\left((n+1)^2 \int_0^{\frac{\pi}{n+1}} t^{\alpha}\, dt\right)$$

$$= O\left(\frac{1}{(n+1)^{\alpha-1}}\right)$$

Now, using Lemma 4.2, we get

$$I_2 = O\left(\int_{\frac{\pi}{n+1}}^{\pi} (t^{\alpha}) \frac{1}{t^2 (n+1)}\, dt\right)$$

$$= O\left[\frac{1}{(n+1)} \int_{\frac{\pi}{n+1}}^{\pi} t^{\alpha-2}\, dt\right]$$

$$= \begin{cases} O\left[\dfrac{1}{(n+1)} \dfrac{1}{(1-\alpha)}\left(\dfrac{\pi^{\alpha-1}}{(n+1)^{\alpha-1}} - \pi^{\alpha-1}\right)\right], & 0<\alpha<1 \\ O\left(\dfrac{\log(n+1)}{(n+1)}\right), & \alpha = 1 \end{cases}$$

$$= \begin{cases} O\left(\dfrac{1}{(n+1)^{\alpha}}\right), & 0<\alpha<1 \\ O\left(\dfrac{\log(n+1)}{(n+1)}\right), & \alpha = 1. \end{cases}$$

Therefore, equation (4.2) becomes

$$\| t_n^{ME} - f \|_{L^r_{(\alpha)}} = O\left(\frac{1}{(n+1)^{\alpha-1}}\right) + \begin{cases} O\left(\frac{1}{(n+1)^{\alpha}}\right), 0<\alpha<1 \\ O\left(\frac{\log(n+1)}{(n+1)}\right), \alpha=1 \end{cases}$$

This completes the proof.

Theorem 4.2. Let (M,λ_n) be a regular method defined by a positive sequence $\{\lambda_n\}$ satisfying conditions (2.1).

Let $f:[0,2\pi]\rightarrow\mathbb{R}$ *be* 2π*-periodic, integrable in the sense of Lebesgue, and belonging to the class* $\text{Lip}(\xi(t),r)$, $r \geq 1$, *then the degree of approximation of* f *by* t_n^{ME} *means of its Fourier series* (1.1) *is given by*

$$\| t_n^{ME} - f \|_{L^r_{(\xi)}} = O\left((n+1)^2 \int_0^{\frac{\pi}{n+1}} \xi(t)\,dt\right) + O\left(\frac{1}{(n+1)} \int_{\frac{\pi}{n+1}}^{\pi} \frac{\xi(t)}{t^2}\,dt\right)$$

Proof. In the proof of Theorem 4.1, by using Lemma 4.4 and generalized Minkowski inequality, we have

$$\begin{aligned}
\| t_n^{ME} - f \|_{L^r_{(\xi)}} &= \left[\int_0^{2\pi} \left|t_n^{ME}(x) - f(x)\right|^r dx\right]^{\frac{1}{r}} \\
&= \left[\int_0^{2\pi} \left|\int_0^{\pi} \varphi(x,t)(ME)_n(t)\,dt\right|^r dx\right]^{\frac{1}{r}} \\
&\leq \int_0^{\pi} \left\{\int_0^{2\pi} |\varphi(x,t)|^r\, dx\right\}^{\frac{1}{r}} |(ME)_n(t)|\,dt \\
&= \int_0^{\pi} O(\xi(t))\,|(ME)_n(t)|\,dt \\
&= O\left(\int_0^{\frac{\pi}{n+1}} (\xi(t))(ME)_n(t)\,dt\right) + O\left(\int_{\frac{\pi}{n+1}}^{\pi} (\xi(t))(ME)_n(t)\,dt\right) \\
&= I_1' + I_2' \text{ (say)}
\end{aligned} \tag{4.6}$$

Now, using Lemma 4.1, we have

$$I_1' = O\left(\int_0^{\frac{\pi}{n+1}} (n+1)^2\, \xi(t)\,dt\right)$$

$$=O\left((n+1)^2\int_0^{\frac{\pi}{n+1}}\xi(t)\,dt\right)$$

Applying Lemma 4.2, we get

$$I_2' = O\left(\frac{1}{(n+1)}\int_{\frac{\pi}{n+1}}^{\pi}\frac{\xi(t)}{t^2}dt\right)$$

Therefore, equation (4.3) becomes

$$\| t_n^{ME} - f \|_{L^r_{(\xi)}} = O\left((n+1)^2\int_0^{\frac{\pi}{n+1}}\xi(t)\,dt\right) + O\left(\frac{1}{(n+1)}\int_{\frac{\pi}{n+1}}^{\pi}\frac{\xi(t)}{t^2}dt\right)$$

This completes the proof.

4.3 APPROXIMATION RESULTS CONCERNING CONJUGATE SERIES OF FOURIER SERIES OF FUNCTIONS

Like the previous section, we begin this section with the following lemmas that will be required to establish the main results of this section.

Lemma 4.1. *For* $0 < t \le \frac{\pi}{n+1}$, $(ME)_n^{\sim}(t) = O\left(\frac{1}{t}\right)$.

Proof. For $0 < t \le \frac{\pi}{n+1}$, $\sin\left(\frac{t}{2}\right) \ge \frac{t}{\pi}$ and $|\cos nt| \le 1$, we have

$$|(ME)_n^{\sim}(t)| = \left|\frac{1}{2\pi}\sum_{k=0}^{n}\lambda_k \frac{\cos^{n-k}\left(\frac{t}{2}\right)\cos\left(n-k+\frac{1}{2}\right)(t)}{\sin\left(\frac{t}{2}\right)}\right|$$

$$\le \frac{1}{2\pi}\sum_{k=0}^{n}|\lambda_k|\frac{\pi}{t}$$

$$= O\left(\frac{1}{t}\right)$$

Lemma 4.2. *For* $\frac{\pi}{n+1} < t \le \pi$, $(ME)_n^{\sim}(t) = O\left(\frac{1}{t^2(n+1)}\right)$ *under the conditions* $(n+1)\lambda_n = O(1)$ *and* $\sum_{k=0}^{n-1}|\Delta\lambda_k| = O\left(\frac{1}{n+1}\right)$.

Proof. Suppose, the conditions $(n+1)\lambda_n = O(1)$ and $\sum_{k=0}^{n-1}|\Delta\lambda_k| = O\left(\frac{1}{n+1}\right)$ hold.

For $\frac{\pi}{n+1} < t \le \pi$, $\sin\left(\frac{t}{2}\right) \ge \frac{t}{\pi}$, $|cos^{n-k}\left(\frac{t}{2}\right)| \le 1$, $0 < k < n$, and using Abel's lemma, we have

$$
\begin{aligned}
|(ME)_n^{\sim}(t)| &= |\frac{1}{2\pi}\sum_{k=0}^{n}\lambda_k \frac{cos^{n-k}\left(\frac{t}{2}\right)\cos\left(n-k+\frac{1}{2}\right)(t)}{\sin\left(\frac{t}{2}\right)}| \\
&\le \frac{1}{2t}\sum_{k=0}^{n}\left|\lambda_k \cos\left(n-k+\frac{1}{2}\right)(t)\right| \\
&\le \frac{1}{2t}\left[\sum_{k=0}^{n-1}|\lambda_k-\lambda_{k+1}|\sum_{j=0}^{k}\cos\left(n-j+\frac{1}{2}\right)(t)+\lambda_n\sum_{k=0}^{n}\cos\left(n-k+\frac{1}{2}\right)(t)\right] \\
&\le \frac{1}{2t}\left[\sum_{k=0}^{n-1}|\Delta\lambda_k|+\lambda_n\right]\max_{0\le k\le n}|\sum_{j=0}^{k}\cos\left(n-j+\frac{1}{2}\right)(t)| \\
&= O\left(\frac{1}{t^2}\right)\left[\sum_{k=0}^{n-1}|\Delta\lambda_k|+\lambda_n\right] \\
&= O\left(\frac{1}{t^2}\right)\left(\frac{1}{n+1}+\frac{1}{n+1}\right) \\
&= O\left(\frac{1}{t^2(n+1)}\right)
\end{aligned}
$$

Lemma 4.3 (see [20]). *If* $f \in \mathrm{Lip}(\alpha, r)$, $0 < \alpha \le 1$, $r \ge 1$, *then*

$$\left[\int_0^{2\pi}|\varphi\sim(x,t)|^r\,dx\right]^{\frac{1}{r}} = O\left(|t|^{\alpha}\right)$$

Lemma 4.4 (see [20]). *If* $f \in \mathrm{Lip}(\xi(t), r)$, $r \ge 1$, *then*

$$\left[\int_0^{2\pi}|\varphi\sim(x,t)|^r\,dx\right]^{\frac{1}{r}} = O(\xi(t))$$

Theorem 4.3. Let (M,λ_n) be a regular method defined by a positive sequence $\{\lambda_n\}$ such that

$$(n+1)\lambda_n = O(1) \text{ and } \sum_{k=0}^{n-1}|\Delta\lambda_k| = O\left(\frac{1}{n+1}\right) \tag{4.7}$$

Let $f:\ [0,2\pi]\to\mathbb{R}$ *be* 2π *-periodic, integrable in the sense of Lebesgue, and belonging to the class Lip*(α,r), $r\geq 1$, *then the degree of approximation of* f *by the* $(M,\lambda_n)(E,1)$ *transform*

$$t_n^{\sim ME} = \sum_{k=0}^{n}\lambda_k \frac{1}{2^{n-k}}\sum_{v=0}^{n-k}\binom{n-k}{v}s_v^{\sim}$$

of its conjugate series (4.2) satisfies

$$\| t_n^{\sim ME} - f^{\sim} \|_{L^r_{(\alpha)}} = \begin{cases} O\left(\dfrac{1}{(n+1)^{\alpha}}\right), 0<\alpha<1 \\ O\left(\dfrac{\log(n+1)}{(n+1)}\right), \alpha = 1. \end{cases} \text{, for } n = 0, 1, 2, 3, \ldots$$

Proof. Let $s_n^{\sim}(f,x)$ denotes the partial sum of series (4.2) and $f^{\sim}$ be the conjugate of the function f ; then we have

$$s_n^{\sim}(f,x) - f^{\sim}(x) = \frac{1}{2\pi}\int_0^{\pi}\varphi\sim(x,t)\frac{\cos\left(n+\frac{1}{2}\right)t}{\sin\left(\frac{t}{2}\right)}dt$$

By $E_n^{\sim(1)}(x)$, we denote $(E,1)$ means of $s_n^{\sim}(f,x)$, and we get

$$\frac{1}{2^n}\sum_{k=0}^{n}\binom{n}{k}\left\{s_k^{\sim}(f,x) - f^{\sim}(x)\right\} = \frac{1}{2^{n+1}\pi}\int_0^{\pi}\frac{\varphi\sim(x,t)}{\sin\left(\frac{t}{2}\right)}\sum_{k=0}^{n}\binom{n}{k}\cos\left(k+\frac{1}{2}\right)tdt$$

$$E_n^{\sim(1)}(x) - f^{\sim}(x) = \frac{1}{2^{n+1}\pi}\int_0^{\pi}\frac{\varphi\sim(x,t)}{\sin\left(\frac{t}{2}\right)} \operatorname{Re}\left\{\sum_{k=0}^{n}\binom{n}{k}e^{i\left(k+\frac{1}{2}\right)t}\right\}dt$$

$$= \frac{1}{2^{n+1}\pi}\int_0^{\pi}\frac{\varphi\sim(x,t)}{\sin\left(\frac{t}{2}\right)} \operatorname{Re}\left\{e^{it/2}\left(1+e^{it}\right)^n\right\}dt$$

$$=\frac{1}{2^{n+1}\pi}\int_0^{\pi}\frac{\varphi\sim(x,t)}{\sin\left(\frac{t}{2}\right)}\text{ Re }\left\{2^n cos^n\left(\frac{t}{2}\right)e^{i(n+1)t/2}\right\}dt$$

$$=\frac{1}{2\pi}\int_0^{\pi}\varphi\sim(x,t)\frac{cos^n\left(\frac{t}{2}\right)\cos(n+1)\left(\frac{t}{2}\right)}{\sin\left(\frac{t}{2}\right)}dt$$

By $t_n^{\sim ME}(x)$, we denote (M,λ_n) means of $E_n^{\sim(1)}(x)$, and we obtain

$$\sum_{k=0}^{n}\lambda_k\left\{E_{n-k}^{\sim(1)}(x)-f^{\sim}(x)\right\}$$

$$=\sum_{k=0}^{n}\lambda_k\left\{\frac{1}{2\pi}\int_0^{\pi}\varphi\sim(x,t)\frac{cos^{n-k}\left(\frac{t}{2}\right)\cos\left(n-k+\frac{1}{2}\right)(t)}{\sin\left(\frac{t}{2}\right)}dt\right\}$$

$$t_n^{\sim ME}(x)-f^{\sim}(x)=\int_0^{\pi}\varphi\sim(x,t)(ME)_n^{\sim}(t)dt$$

Using Lemma 4.3 [26] (generalized Minkowski inequality), we have

$$\|t_n^{\sim ME}-f^{\sim}\|_{L^r_{(\alpha)}}=\left[\int_0^{2\pi}\left|t_n^{\sim ME}(x)-\tilde{f}(x)\right|^r dx\right]^{\frac{1}{r}}$$

$$=\left[\int_0^{2\pi}\left|\int_0^{\pi}\varphi\sim(x,t)(ME)_{\tilde{n}}(t)dt\right|^r dx\right]^{\frac{1}{r}}$$

$$\leq\int_0^{\pi}\left\{\int_0^{2\pi}|\varphi\sim(x,t)|^r dx\right\}^{\frac{1}{r}}|(ME)_{\tilde{n}}(t)|dt$$

$$=\int_0^{\pi}O\left(t^{\alpha}\right)|(ME)_{\tilde{n}}(t)|dt$$

$$=O\left(\int_0^{\frac{\pi}{n+1}}(t^{\alpha})(ME)_{\tilde{n}}(t)dt\right)+O\left(\int_{\frac{\pi}{n+1}}^{\pi}(t^{\alpha})(ME)_{\tilde{n}}(t)dt\right)$$

$$=I_1+I_2\text{, say}\tag{4.8}$$

Applying Lemma 4.1, we have

$$I_1 = O\left(\int_0^{\frac{\pi}{n+1}} (t^{\alpha}) \frac{1}{t} dt\right)$$

$$= O\left(\frac{1}{(n+1)^{\alpha}}\right)$$

Now, using Lemma 4.2, we get

$$I_2 = O\left(\int_{\frac{\pi}{n+1}}^{\pi} (t^{\alpha}) \frac{1}{t^2 (n+1)} dt\right)$$

$$= O\left[\frac{1}{(n+1)} \int_{\frac{\pi}{n+1}}^{\pi} t^{\alpha-2} dt\right]$$

$$= \begin{cases} O\left[\frac{1}{(n+1)} \frac{1}{(1-\alpha)} \left(\frac{\pi^{\alpha-1}}{(n+1)^{\alpha-1}} - \pi^{\alpha-1}\right)\right], 0 < \alpha < 1 \\ O\left(\frac{\log(n+1)}{(n+1)}\right), \alpha = 1 \end{cases}$$

$$= \begin{cases} O\left(\frac{1}{(n+1)^{\alpha}}\right), 0 < \alpha < 1 \\ O\left(\frac{\log(n+1)}{(n+1)}\right), \alpha = 1 \end{cases}$$

Therefore, from (4.2) we have

$$\| t_n^{\sim ME} - f^{\sim} \|_{L^r_{(\alpha)}} = \begin{cases} O\left(\frac{1}{(n+1)^{\alpha}}\right), 0 < \alpha < 1 \\ O\left(\frac{\log(n+1)}{(n+1)}\right), \alpha = 1 \end{cases}$$

for $n = 0, 1, 2, 3, \ldots$

This completes the proof.

Theorem 4.4. Let (M, λ_n) *be a regular method defined by a positive sequence* $\{\lambda_n\}$ *satisfying conditions (4.6).*

Let $f : [0,2\pi] \to \mathbb{R}$ *be* 2π*-periodic, integrable in the sense of Lebesgue, and belonging to the class* $Lip(\xi(t),r)$, $r \geq 1$*; then the degree of approximation of* f *by* $t_n^{\sim ME}$ *means of its conjugate series (4.2) is given by*

$$\| t_n^{\sim ME} - f^{\sim} \|_{L^r_{(\xi)}} = O\left(\int_0^{\frac{\pi}{n+1}} \frac{\xi(t)}{t} dt \right) + O\left(\frac{1}{(n+1)} \int_{\frac{\pi}{n+1}}^{\pi} \frac{\xi(t)}{t^2} dt \right) \tag{4.9}$$

Proof. In the proof of Theorem 4.1, using Lemma 4.4 and generalized Minkowski inequality, we get

$$\begin{aligned}
\| t_n^{\sim ME} - f^{\sim} \|_{L^r_{(\xi)}} &= \left[\int_0^{2\pi} \left| t_n^{\sim ME}(x) - f^{\sim}(x) \right|^r dx \right]^{\frac{1}{r}} \\
&= \left[\int_0^{2\pi} \left| \int_0^{\pi} \varphi \sim (x,t) (ME)_n^{\sim}(t) dt \right|^r dx \right]^{\frac{1}{r}} \\
&\leq \int_0^{\pi} \left\{ \int_0^{2\pi} | \varphi \sim (x,t) |^r dx \right\}^{\frac{1}{r}} |(ME)_n^{\sim}(t)| dt \\
&= \int_0^{\pi} O(\xi(t)) |(ME)_n^{\sim}(t)| dt \\
&= O\left(\int_0^{\frac{\pi}{n+1}} (\xi(t)) (ME)_n^{\sim}(t) dt \right) + O\left(\int_{\frac{\pi}{n+1}}^{\pi} (\xi(t)) (ME)_n^{\sim}(t) dt \right) \\
&= I_1' + I_2' \quad \text{(say)}
\end{aligned} \tag{4.10}$$

Now using Lemma 4.1, we have

$$I_1' = O\left(\int_0^{\frac{\pi}{n+1}} \frac{\xi(t)}{t} dt \right)$$

Applying Lemma 4.2, we get

$$I_2' = O\left(\frac{1}{(n+1)} \int_{\frac{\pi}{n+1}}^{\pi} \frac{\xi(t)}{t^2} dt \right)$$

Therefore, equation (4.9) becomes

$$\| t_n^{\sim ME} - f^{\sim} \|_{L^r_{(\xi)}} = O\left(\int_0^{\frac{\pi}{n+1}} \frac{\xi(t)}{t} dt \right) + O\left(\frac{1}{(n+1)} \int_{\frac{\pi}{n+1}}^{\pi} \frac{\xi(t)}{t^2} dt \right)$$

Example 4.1. *If we take* $\lambda_k = \dfrac{1}{\sqrt{n^2+k}}$, $k = 1,2,3,\ldots$, $\lambda_0 = 0$, *then the degree of approximation of a function f belonging to the classes* $Lip(\alpha,r)$ *and* $Lip(\xi(t),r)$, $r \geq 1$ *by* $(M,\lambda_n)(E,1)$ *means*

$$t_n^{\sim ME} = \sum_{k=0}^{n} \frac{1}{\sqrt{n^2+k}} \frac{1}{2^{n-k}} \sum_{v=0}^{n-k} \binom{n-k}{v} s_v^{\sim}$$

of its conjugate series (4.2) are given by

$$\| t_n^{\sim ME} - f^{\sim} \|_{L^r_{(\alpha)}} = \begin{cases} O\left(\dfrac{1}{(n+1)^{\alpha}}\right), 0<\alpha<1 \\ O\left(\dfrac{\log(n+1)}{(n+1)}\right), \alpha = 1 \end{cases}$$

for $n = 0, 1, 2, 3, \ldots$
and

$$\| t_n^{\sim ME} - f^{\sim} \|_{L^r_{(\xi)}} = O\left(\int_0^{\frac{\pi}{n+1}} \frac{\xi(t)}{t} dt\right) + O\left(\frac{1}{(n+1)} \int_{\frac{\pi}{n+1}}^{\pi} \frac{\xi(t)}{t^2} dt\right), \text{respectively.}$$

In a similar fashion, we can also establish the following result (see also [20]).

Theorem 4.5. *Let* (M,λ_n) *be a regular method defined by a positive sequence* $\{\lambda_n\}$ *satisfying conditions (4.6). Let* $\xi(t)$ *be a modulus of continuity such that*

$$\int_0^{v} \frac{\xi(t)}{t} dt = O(\xi(v)), 0 < v < \pi$$

Let $f: [0,2\pi] \to \mathbb{R}$ *be* 2π *-periodic, integrable in the sense of Lebesgue, and belonging to the class* $Lip(\xi(t),r)$, $r \geq 1$; *then the degree of approximation (4.8) of* f *by* $t_n^{\sim ME}$ *means of its conjugate series (4.2) becomes*

$$\| t_n^{\sim ME} - f^{\sim} \|_{L^r_{(\xi)}} = O\left(\frac{1}{(n+1)} \int_{\frac{\pi}{n+1}}^{\pi} \frac{\xi(t)}{t^2} dt\right) \tag{4.11}$$

In addition, if $\left(\dfrac{\xi(t)}{t}\right)$ *is monotonically decreasing in* $\left(\dfrac{\pi}{n+1}, \pi\right)$, *then (4.10) reduces to*

$$\| t_n^{\sim ME} - f^{\sim} \|_{L^r_{(\xi)}} = O\left(\xi\left(\frac{1}{(n+1)}\right) \log(n+1)\right)$$

4.4 CONCLUSION

There are many ways to introduce advanced product summability methods for computing the degree and error of approximation of several other functions (signals). The chapter is expected to be helpful in constructing other product summability methods for applications in approximation theory and signal processing.

4.5 ACKNOWLEDGMENTS

The first draft of the third section was presented at ICAEESE19 held at Gauhati University, India. The first author is supported under the DST-INSPIRE Fellowship scheme with No: IF170072.

REFERENCES

[1] P. Chandra, Trigonometric approximation of functions in L_p norm, *J. Math. Anal. Appl.*, 275(1) (2002), 13–26.

[2] P. Chandra, Approximation by Nörlund operators, *Mat. Vestnik*, 38 (1986), 263–269.

[3] R. N. Mohapatra and D. C. Russell, Some direct and inverse theorems in approximation of functions, *J. Aust. Math. Soc. (Ser. A)*, 34 (1983), 143–154.

[4] G. Alexits, Convergence problems of orthogonal series, translated from German by I, *Földer. Int. Ser. Monogr. Pure Appl. Math.*, 20 (1961).

[5] B. N. Sahney and D. S. Goel, On the degree of continuous functions, *Ranchi Univ. Math. J.*, 4 (1973), 50–53.

[6] B. N. Sahney and V. Rao, Errors bounds in the approximation of functions, *Bull. Australian Math. Soc.*, 6 (1972), 11–18.

[7] B. E. Rhodes, On the degree of approximation of functions belonging to Lipschitz class by Hausdorff means of its Fourier series, *Tamkang J. Math.*, 34(3) (2003), 245–247.

[8] H. H. Khan, On the degree of approximation of a function belonging to the class Lip *(α, p)*, *Indian J. Pure Appl. Math.*, 5 (1974), 132–136.

[9] H. H. Khan, On the degree of approximation to a function belonging $W\left(L_p,\xi(t)\right)$ to weighted class, *Aligarh Bull. Math.*, 3–4 (1973–1974), 83–88.

[10] H. H. Khan, A note on a theorem of Izumi, *Commun. Fac. Maths. Ankara*, 31 (1982), 123–127.

[11] V. N. Mishra and L. N. Mishra, Trigonometric approximation of signals (functions) in L_p-norm, *Int. J. Contemp. Math. Sci.*, 7(12) (2012), 909–918.

[12] K. Qureshi, On the degree of approximation of a periodic function *f* by almost Nörlund means, *Tamkang J. Math.*, 12(1) (1981), 35–38.

[13] K. Qureshi and H. K. Neha, A class of functions and their degree of approximation, *Ganita.*, 41(1) (1990), 37–42.

[14] S. Lal and P. N. Singh, Degree of approximation of conjugate of Lip(α,p) function by $(C,1)(E,1)$ means of conjugate series of a Fourier series, *Tamkang J. Math.*, 33(1) (2002), 269–274.

[15] H. K. Nigam, Approximation of conjugate of a function belonging to Lip$(\xi(t),r)$ class by $(C,1)(E,1)$ product means of conjugate series of Fourier series, *Ultra Sci. Phys. Sci.*, 22(1(M)) (2010), 295–302.

[16] H. K. Nigam, On $(C,1)(E,1)$ product means of Fourier series and its conjugate Fourier series, *Ultra Sci. Phys. Sci.*, 22(2) (2010), 419–428.

[17] H. K. Nigam and K. Sharma, Degree of approximation of a class of function by $(C,1)(E,q)$ means of Fourier series, *IAENG Int. J. Appl. Math.*, 41(2) (2011), 1–5.
[18] S. Lal and A. Mishra, Approximation of signals of generalized Lipschitz class by $(N,p_n)(E,1)$ summability means of Fourier series, *Pacific J. Appl. Math.*, 5(1) (2013), 69–80.
[19] S. Das and H. Dutta, Degree of approximation of signals in certain Lipschitz classes by Zweier-Euler product summability method of Fourier series, *Math. Meth. Appl. Sci.*, 44 (2021), 4881–4887.
[20] V. N. Mishra and V. Sonavane, Approximation of functions of Lipschitz class by $(N,p_n)(E,1)$ summability means of conjugate series of Fourier series, *J. Classical Anal.*, 6(2) (2015), 137–151.
[21] A. Zygmund, *Trigonometric Series*, first ed., New York: Cambridge University Press, 1939.
[22] A. Aasma, H. Dutta and P. N. Natarajan, *An Introductory Course in Summability Theory*, Hoboken: John Wiley & Sons, Inc., 2017.
[23] G. H. Hardy, *Divergent Series*, first ed., Oxford University Press, 1949.
[24] S. Lal and A. Mishra, Approximation of functions of class $\text{Lip}(\alpha,r)$, ($r\geq1$), by $(N,p_n)(E,1)$ summability means of Fourier series, *Tamkang J. Math.*, 45(3) (2014), 243–250.
[25] E. C. Titchmarsh, *The Theory of Functions*, second ed., Oxford University Press, 1939.
[26] A. Zygmund, *Trigonometric Series*, second ed., New York: Cambridge University Press, 1968.

5 NTRU Cryptosystem over Rational Numbers Q_P and a New Verification Method over Rational Functions Field

Mehmet Sever

5.1 INTRODUCTION

Number Theorists 'R' Us (NTRU) was first discussed by Hoffstein, Pipher, and Silverman at Crypto' 96 [1]. Later on, many additions and contributions were made to the system by the developers of NTRU, which is seen as a ring-based and public key encryption method. [2]. They introduced $NTRU_{SIGN}$ in 2003 [3], a digital signature version of NTRU. Also, in the same year, they made a presentation which analyzed decryption errors of NTRU with another team [4]. Silverman published a report about invertible polynomials in a ring in 2003 [5]. Also in 2005, Silverman and Whyte published a technical report which analyzed and discussed error probabilities in the NTRU decryption phase [6]. They placed the system on a mathematical basis in the next years by constantly publishing articles and studies related to the system on the website www.ntru.com.

The system stands out for being highly resistant to quantum-based attacks as well as its speed. The basic reason for protecting this resistance is based on finding a lattice vector with the shortest length and the complexity of the problem of finding a lattice point closest to the private key in a high-dimensional lattice [8]. Unlike other systems, it will be resistant to attacks for a long time because it is based on high-dimensional lattices. The system is being developed day by day by considering different algebraic structures.

Some of the attacks made and the new parameters integrated into the system can be looked at in a work by Coppersmith et al. from 1997 [9]. Subsequently, a richer and more powerful version was re-released, and the system was updated by Hoffstein et al. in 2003 [10].

For another attack on the cryptosystem, the following study can also be examined [11]; It has gained increased importance today by introducing more powerful, current parameters and solutions to the NTRU cryptosystem, which addresses attacks such as the difference-splitting attack [12].

DOI: 10.1201/9781003322856-5

For detailed readings, these can be seen in [13–15] for different types of attacks, and on the contrary, they can be seen in [16–20] for proposed new parameters and new systems.

5.2 AIM AND SCOPE

In this chapter, which aims to carry the NTRU cryptosystem on robust algebraic structures, some interesting properties and results were added to the cryptosystem theoretically. Taking advantage of the fact that the rational numbers field is larger and more complex than a integer ring, more attention has been paid to security, which is the main purpose of cryptology. For this purpose, we have tried to present the newly proposed cryptosystem in a more complex and powerful form. But, at the same time, since rational numbers are determined by two integers with a basic rule, the new proposed system is also considered practical and useful. In light of this chapter, new lattice types will be determined and security analyses can be done by arranging attacks on the proposed NTRU cryptosystem.

5.3 CONTRIBUTIONS AND MAIN RESULTS

The contributions and results that can be summarized as follows are obtained in this chapter:

1. The NTRU type of cryptosystem is recognized for the first time on rational numbers.
2. A field-based approximation method to the NTRU system, which is based on a richer algebraic background in the field of rational numbers Q, is tried for the first time.
3. Plaintext space is enlarged with a detailed architecture in the new structure, and the security degree of the system is increased one more step.
4. In addition, theorems about how to find private keys are given, and the key generation capacity of the system is increased.
5. Also the system is handled and recognized on the rational functions field, which is much more complex than the polynomial ring.
6. Since the rational functions field is a one-dimensional vector space on itself, the system is handled in a new structure with the effect of a base-changing operator to message elements.
7. A new digital signature and authentication method is introduced.
8. The proposed new system contains the classical NTRU lattice, which $2N \times 2N$ in size, as a $4N \times 4N$ lattice-based scheme. Therefore, the system over the rational functions field is more protected against lattice reduction algorithm-based attacks.

5.4 NTRU PARAMETERS

The following parameters will be provided later for use in the encryption and decryption operations of NTRU, as well as in the key generation process:

- N: it determines a maximum degree of polynomials being used. N is chosen as a prime so that the process is preserved against attacks. Also, it is chosen to be big enough that the process is preserved from lattice attacks.
- q: it is a large module of a dual module, and it is chosen as a positive integer. Its values vary depending on the specific goals of the process.
- p: it is a small module of a dual module and generally a positive integer. Also, it is rarely chosen as a polynomial with small coefficients.

The parameters N, q, and p can be differently chosen from the sets according to the preferred security level. The case $(p,q)=1$ is always preserved so that the ideal (p,q) is equal to the whole ring.

- L_f, L_g: sets of private keys in which polynomials are chosen to be kept confidential for encryption.
- L_m: it is a plain or sender's text set. It is stated to be unencrypted and codable as polynomials.
- L_r: it is a set of error polynomials. It is stated to be a set of arbitrarily chosen error polynomials with small coefficients in the phase of encryption.
- *center*: a centralization or facilitator method. An algorithm guaranteeing which *modq* reductions are applied in perfect truth in the phase of decryption.

See [1] for more details of the NTRU parameters which were introduced previously.

5.5 ALGEBRAIC BACKGROUND OF NTRU CRYPTOSYSTEM

5.5.1 Explanations and Notations

The encryption phase of NTRU is performed in a quotient ring $R = Z[x]/(x^N - 1)$. N is a positive and large integer. It is generally chosen as a big prime for not being easily predictable. For $f(x)$, which is a polynomial in R, f_k denotes a coefficient of x_k for every $k \in [0, N-1]$, and $f(x)$ denotes a value of f in x for $x \in \mathbb{C}$. A convolution product $h = f \star g$ is given by $h_k = \sum_{i+j \equiv k mod N} f_i \cdot g_j$, where f and g are two polynomials in R. p and q were chosen as powers of 3 and 2, respectively, when the NTRU system was first introduced. The subset L_m: consists of polynomials such that the coefficients $\{-1,0,1\}$ are called ternary polynomials. The secret key $f \in L_f$ was usually chosen in the form $1 + p \cdot F$ because of easy computing. Recent studies indicate that can be chosen not only as an integer but also as a polynomial.

5.5.2 Phase of Key Generation

1 $f \in L_f$ and $g \in L_g$ are randomly chosen such that f is invertible both in *modp* and *modq*.

2 $F_q \equiv f^{-1} modq$ and $F_p \equiv f^{-1} modp$.
3 A dual private key is (p, F_p).
4 For easy and fast operations in the system, the public key is obtained in the form $H \equiv p \cdot g \star F_q modq$.

It is noted that g cannot be used as a private key in the phase of decryption. Since $H \star f \equiv p \cdot gmodq$, $H \star f \equiv 0modp$ if the private key is changed, then modulo parameters also must be changed, which cannot be used when $modp$ is substituted.

5.5.3 Encryption Phase

In an algorithmic language, the encryption phase is represented;
Input: a message $m \in L_m$ sent and again determined by the sender a public key H.
Output: ciphered message $e \in \Upsilon(m)$

1 Choose $r \in L_r$ randomly.
2 Compute and return $e \equiv r \star H + mmodq$.

Since not all fit, the set $\Upsilon(m)$ denotes plain texts m which can be encrypted easily and fastly.

5.5.4 Phase of Decryption

Again, if the phase of decryption is represented as algorithmic stream, an algorithm D takes e as follows:

Input: a ciphered message $e \in \Upsilon(m)$ and a dual private key (p, F_p).
Output: a plain or decrypted text $D(e) \equiv m \in L_m$.

1 Calculate and obtain $amodq \equiv e \star fmodq$.
2 Compute a polynomial $amodq$ with integer coefficients from $a \equiv p \cdot r \star g + f \star m \in R$ by performing a centralization operation.
3 $mmodp \equiv a \star F_p modp$.
4 a possible plain text $m \equiv \Psi modp$.

It should be noted that Ψ is the mapping $\Psi : m \mapsto mmodp$. That is, it performs as a function $\Psi : L_m \to L_m modp$. It is very important to choose convenient parameters in order to work decryption operations completely, that is, $D(e) = m$.

5.6 ALGEBRAIC BACKGROUND OF PROPOSED NEW NTRU SYSTEM

All parameters will be faithfully used in the classic $R_p = Z_p[X]/(X^N - 1)$ NTRU, ring but only polynomials will be selected over Q_p. The convolution $(*)$ product is similarly defined for $f, g \in Q_p[X]/(X^N - 1)$ such that

$$f*g=\sum_{i+j\equiv K(modN)} f_i g_j x^{i+j}$$

$(p,q)=1$ is being $p<q$ also applies here. Thus, in the newly proposed system $f,g\in Q_p[X]/(X^N-1)$ is it, f_q^{-1} and f_p^{-1} are available r error polynomial is chosen arbitrarily and for selected message polynomial m to be sent,

$$e\equiv ph*r+m(modq)$$

to be obtained in the form of an e encryption polynomial. Thus, a new cryptosystem on Q_p will be considered.

Remark 5.1. *Thanks to this system and since Q_p contains Z_p, the public key h and message polynomial m spaces will grow much larger, and a more protected system will be obtained against attacks.*

Remark 5.2. *Also in the ring $Q_p[X]/(X^N-1)$, the $(f,X^N-1)=1$ condition is necessary and sufficient for the private key f selections.*

Now, let's give a lemma that allows operations to be performed in Z_p instead of Q_p if necessary.

Lemma 5.1. *The product of the polynomial $h\equiv f_q^{-1}*g(modq)$ and the arbitrarily chosen R error polynomial is movable to Z_p.*

Proof. Where f_q^{-1} and g are members of $Q_p[X]/(X^N-1)$, the product of these polynomials is equal to $h\in Q_p[X]/(X^N-1)$. Thus, the h polynomial is reducible over Q_p. So, it must be reducible on Z_p from the Gauss lemma. That is, for appropriately selected $r,s\in Q$, rf_q^{-1} and sg can be made elements of $Z[X]$:

if $f_q^{-1}=a_nx^n+\cdots+a_1x+a_0$ and $g=b_nx^n+\cdots+b_1x+b_0$, then, for arbitrarily selected i,j,k, it would be

$$\begin{aligned} rsh=rs\sum_{i+j\equiv K(modN)} a_ib_jx^k &\equiv r\sum a_ix^i s\sum b_jx^j\left(modX^N-1\right)\\ &\equiv rsf_q^{-1}g\left(modX^N-1\right)\\ &\equiv rf_q^{-1}sg\left(modX^N-1\right)\end{aligned}$$

from here it can be seen that the rsh polynomial is an element of $Z[X]$. Similarly, the R error polynomial can be written,

$$R=r'R'$$

for $R'\in Z[X]$ and $r'\in Q$.

Thus, for randomly selected h polynomials, there is an R error polynomial such that $h*R$ is in $Z_p[X]$.

Remark 5.3. *These selections of R error polynomials allow one to make operations in* $Z[X]$ *and therefore increase the speed of the system.*

5.7 SET OF MESSAGEABLE POLYNOMIALS

Every polynomial in the $R_p = Q_p[X]/(X^N - 1)$ ring is actually messageable. The only criterion to be considered is that a polynomial of type $m = \sum \alpha_i x^i$, $\alpha_i = \frac{r_i}{s_i} \in Q$, if $(r_i, s_i) = 1$, can be deciphered improbably. But if this condition is not provided for at least one $\alpha_j = \frac{r_j}{s_j}$ and $(r_j, s_j) \neq 1$, then the numbers are taken as such: $\frac{r_j}{s_j} = \frac{h_j}{k_j}, (h_j, k_j) = 1, h_j, k_j \in Z$. In this case, these two polynomials with coefficients $\alpha_j = \frac{r_j}{s_j}$ and $\alpha_j = \frac{h_j}{k_j}$ can be equated as a code word since they are equal on $Q_p[X]$. Now, on this assumption, a bijective transformation between the message polynomials and code words can be given.

Theorem 5.1. *For* $\alpha_i = \frac{r_i}{s_i} \in Q$, *the* τ *function defined as*

$$\tau : Q_p[X]/(X^N - 1) \to Q_p^N \cong Z_p^{2N}$$

$$\tau\left(p(x) = \sum \alpha_i x^i\right) = (\alpha_1, \alpha_2, \ldots, \alpha_N) \cong (r_1, s_1, r_2, s_2 \ldots, r_n, s_n)$$

is well-defined and bijectional.

Proof. For two polynomials $P_1(x) = P_2(x)$, but if at least one $\alpha_i = \frac{r_j}{s_j} = \frac{h_j}{k_j}$, $(h_j, k_j) = 1$, then these are actually equal, but only the coefficients are not coprime in between. Therefore, polynomials that are assumed to have equal code words are also equal.

At least one coefficient corresponding to the same index of two polynomials such as $P_1(x) \neq P_2(x)$ is different. Therefor, the 2_i and 2_{i+1}th terms of codewords are different form other. Thus, different codewords will be obtained.

If a sequence in the form of $a \in Z_p^{2N}$ and $a = (a_1, a_2, a_3, a_4, \ldots, a_{2N-1}, a_{2N})$ is determined, then the polynomial $a(x)$ is uniquely determined such that:

$$a(x) = \frac{a_1}{a_2} + \frac{a_3}{a_4} x + \cdots + \frac{a_{2N-1}}{a_{2N}} x^N$$

Remark 5.4. *The message word can also be considered* $\alpha = (\alpha_1, \alpha_2, \ldots, \alpha_N)$. *But if it is accepted as* $\alpha = (r_1, s_1, r_2, s_2 \ldots, r_n, s_n)$ *and sent, a* $2N$ *long word will be added to the classic NTRU cryptosystem so it looks like two messages have been sent at the same time.*

Remark 5.5. *If the sender has not paid attention to the fact that the numerator and denominator forming the coefficients of the message polynomial are coprime,* *then the receiver does not. Mathematically, for the jth term of* $P(x)$, $\alpha_j = \frac{r_i}{s_i}$, *and* $(r_i, s_i) \neq 1$, *the receiver obtains that the code word is*

$$(\ldots, r_i, s_i, \ldots)$$

Remark 5.6. *It is important not to use repetitive numbers in code words to protect against plaintext attacks.*

5.8 SELECTION AND CONSTRUCTION OF INVERTIBLE (PRIVATE KEY) POLYNOMIALS

Since the ring $Q_p[X]/(X^N - 1)$ is larger than the ring $Z_p[X]/(X^N - 1)$, the number of polynomials that can be inverted is greater. Therefore, the ring $Q_p[X]/(X^N - 1)$ is more protected against private key attacks. However, since this resistance will extend private key selections during the use of the system, an algorithm will be given to shorten the process and operations.

Theorem 5.2. *Let* $f \in Q_p[X]/(X^N - 1)$, $f = \sum \alpha_i x^i$, $\alpha_i = \frac{r_i}{s_i}$, *and* $(f, X^N - 1) = 1$. *Let's denote the least common multiple of the numbers in the denominator with* S. *That is, LCM* $(s_1, s_2, \ldots, s_n) = S$. *If* S^{-1} *is the inverse of* S *in modula* p *and if* $(\alpha, p) = (\alpha, S^{-1}) = 1$, *then the* f *polynomial is invertible if and only if there is a* $k \in Z$ *such that* $S^{-1} = \alpha(kp + 1)$.

Proof. In order to reduce operations to Z_p, let us consider the $g = Sf\,(mod p)$ polynomial. It is obvious that $g \in Z_p[X]/(X^N - 1)$. Because of $(f, X^N - 1) = (g, X^N - 1) = 1$, g is invertible. That is, for at least one $g^{-1} \in Z_p[X]/(X^N - 1)$, $g^* g = 1\,(mod\ p)$ is provided. Thus, for every $\alpha \neq 1$ and $\alpha \in Z_p$, the equation

$$(\alpha g) * \left(\frac{1}{\alpha} g\right) \equiv 1\,(mod\, p)$$

can also be considered a equation in $Q_p[X]$. Since

$$\frac{1}{\alpha} g' = \frac{1}{\alpha}(Sf)^{-1} = \frac{1}{\alpha} S^{-1} f^{-1}$$

provided

$$\left(\frac{1}{\alpha} S^{-1} f^{-1}\right) * f \equiv 1(modp)$$

must be $\frac{1}{\alpha} S^{-1} \equiv 1(modp)$. That is, $S^{-1} = \alpha(kp + 1)$.

Remark 5.7. *Theorem 5.2 guarantees the existence of many invertible polynomials for suitable α, k, and p parameters. Also, it is sufficient to look at the denominators in the coefficients of polynomials that can be produced in this way.*

Lemma 5.2. *Let C be the set of all separable and monic polynomials in $Q_p[X]/(X^N-1)$. Also let D be the subset of Q which contains all the roots of the polynomials in C. A φ function, defined as*

$$\varphi : C \to D, \varphi\left(f = \Sigma \alpha_i x^i\right) = \{\beta_1, \beta_2, ..., \beta_n \mid f(\beta_i) = 0\}$$

is well-defined and bijectional. Also, it can be added to the new proposed system as a private key.

Proof. For every $f \in C$, the representation $f = (\alpha_1, \alpha_2, ..., \alpha_n)$ is unique. Also, since $f = (\beta_1 - x)(\beta_2 - x)...(\beta_n - x)$, each set of $\{\beta_i\}$ gives a different f polynomial. Since $\varphi(m)$ is a set, let $\varphi\ (m)$ be a corresponding polynomial. The corresponding polynomial $\varphi\ (m)$ will be added to the proposed cryptosystem as follows;

$$e \equiv pr^*h + \varphi\ (m)\ (mod\ q)$$

After the required decryption steps are done, as in the classical NTRU system,

$$e \equiv \varphi\ (m)(modp)$$

comes to the stage. Now, if $\varphi(m) = \{\beta_1, \beta_2, ..., \beta_n\}$, then $\varphi\ (m) = \Sigma \beta_i x^i$ is it. After all these steps, the receiver solves the equation

$$(x - \beta_1)(x - \beta_2)...(x - \beta_n) = (\alpha_1, \alpha_2, ..., \alpha_n) = m$$

and because he/she knows the inverse of the φ function, this final step is very easy.

Remark 5.8. *This lemma is also applied when the roots are not in $Q_p[X]/(X^N-1)$ but in $Q_{p^n}[X]/(X^N-1)$, which is the expansion field. From the finite fields theory, if $\alpha \in Q_{p^n}$ is a root, then the other roots are powers of α. Thus, the root set is $\left\{\alpha, \alpha^p, ..., \alpha^{p^{n-1}}\right\}$. In this case, when only one root is found, the others are found quickly, and the system speeds up.*

Lemma 5.3. *Let's define f as*

$$f : Q_p[X]/(X^N-1) \to Q_p[X]/(X^N-1), f(m) = m^p\ (modp)$$

Then, f is one to one and can be used in the proposed system.

Proof. Since $Q_p[X]$ is a finite field which has characteristic p, then for $m \in Q_p[X]$, $m^p(X) = m(X^p)$ will satisfy β_i as a root of m. So

$$m^p(\beta_i) = m(\beta_i^p)$$

and from this we get that β_i^p is the other root of m. Thus, a permutation of the roots of the polynomial m will be sent.

Now, let us look at a lemma that allows one more degree of hiding the message in the system.

Lemma 5.4. *Let $m(x) \in Q_p[X]/(X^N - 1)$ be an odd-order polynomial for all powers. Also, let m compute the property that all roots of m are different and monic polynomials. Under such a choice of m, the following equation is possible to use in the system.*

$$e \equiv ph{*}r + m(m(x)) + m(x)(mod\ q) \tag{5.1}$$

Proof. Let β_i s be different roots of $m(x)$ for $1 \le i \le n$. Since the degree of m is odd, it is onto. Thus, for every β_i, there is only one $x_i \in Q$ such that $m(x_i) = \beta_i$. Because if

$$x_i = x_j \Rightarrow \beta_i = m(x_i) = m(x_j) = \beta_j$$

it would be. Therefore, since β_i s are different, x_i s are also different. Now, decryption processes are continued as in the classical method; then Equation 5.1 will be

$$e \equiv m(m(x)) + m(x)(modp) \tag{5.2}$$

where $m(m(x)) = 0$, and the receiver knows $\{x_i\}$ or this set of x_i arrives from a secure channel. Thus, the receiver applies this set of x_i to Equation 5.2,

$$e \equiv 0 + m(x)(modp)$$

Since the receiver knows that $\{x_i\}$ arrives to the $\{m(x_i) = \beta_i\}$, then the receiver will easily identify the polynomial whose roots are found.

Now, with the help of the following lemma, an important method of searching for private key f polynomials in the proposed cryptosystem will be given.

Lemma 5.5. *Let f be a primitive polynomial in $Z_p[X]/(X^N - 1)$. Then, f is irreducible in $Z_p[X]/(X^N - 1)$ if and only if f is irreducible in $Q_p[X]/(X^N - 1)$.*

Proof. Let f be reducible in $Q_p[X]$. Then, there are g,h in $Q_p[X]$ such that $f = gh$. Therefore, also $f = cg_0h_0$ in $Q_p[X]$ for g_0, h_0 primitive polynomials. Hence, there is a $c \in Z^+$ such that,

$$c\,f_{0,} = f = cg_0h_0$$

But this contradicts the irreducibility of f in $Z_p[X]/(X^N-1)$. Therefore, f is irreducible in $Q_p[X]/(X^N-1)$. Because an irreducible f polynomial also guarantees $(f,X^N-1)=1$, a private key f is obtained.

Now, with the help of the given lemma, once a private key is found, the method for deriving the other key from the first will be explained.

Lemma 5.6. *Let $f(x)=f_nx^n+f_{n-1}x^{n-1}+\cdots+f_1x+f_0$ be a nonzero coefficient polynomial in $Q_p[x]$. Using the same coefficients, let a g polynomial be determined by $g(x)=f_0x^n+\cdots+f_{n-1}x+f_n$. Then f being irreducible in $Q_p[x]$ if and only if g is irreducible in $Q_p[x]$ will be realized.*

Proof. Let f be reducible. Then there are $m_1,m_2\in Q_p[x]$ such that $f(x)=m_1(x)m_2(x)$. If $degm_1=i$ and $degm_2=j$ then $i+j=n$ will be true. If $g(x)=f(x^{-1})x^n=a(x^{-1})x^ib(x^{-1})x^j$, then g is reducible in $Q_p[x]$.

5.9 NTRU CRYPTOSYSTEMS OVER THE RATIONAL FUNCTIONS FIELD

This section discusses the new NTRU cryptosystem in the field of rational functions and gives some new generalizations. For any two polynomials p and q, a rational function is type of $\frac{p(x)}{q(x)}$. Also, the set of rational functions, according to ordinary polynomial operations, generates a field structure. But now, a different set of rational functions and a different operation will be given. Therefore, a new field structure will be created. Let $p(x)$ and $q(x)$ be two monic coprime and primitive polynomials in $Q_p[X]/(X^N-1)$. Let $k(x)$ denote the $\frac{p(x)}{q(x)}$ expressions that will occur under these selections of $p(x)$ and $q(x)$. That is, $\frac{p(x)}{q(x)}=k(x)$ will be true.

Let $K(x)$ denote all the choices of $k(x)$ in this way. Now let's give a new multiplication operation on $K(x)$ as follows by using the convolution product $(*)$ in the classical NTRU ring. This operation represented by $\circledast$ is as follows: for any two $k_1(x),k_2(x)\in K(x)$, the $\circledast$ operation is:

$$k_1(x)\circledast k_2(x)=\frac{p_1(x)}{q_1(x)}\circledast\frac{p_2(x)}{q_2(x)}=\frac{p_1(x)*p_2(x)}{q_1(x)*q_2(x)}$$

It is obvious that the $\circledast$ operation is well defined and $K(x)$ is the field according to the $\circledast$ operation and ordinary rational function addition. Now, the encryption part of the NTRU type system that will be on $K(x)$ will given as:

for $f \circledast f^{-1} \equiv 1(modp)$ and $f \circledast f^{-1} \equiv 1(modq)$, f as a private key element and $h = f_q^{-1} \circledast g$ as a public key element are chosen. The encryption algorithm is as follows:

$$e \equiv pr \circledast h + m(modq) \tag{5.3}$$

Remark 5.9. *It is clear that every element* f *is in the form* $f = \dfrac{p_1(x)}{q_1(x)} \in K(x)$, *and for* $p_1(x) \neq 1$. f *is a private key because* $f * \dfrac{1}{f} = 1$. *But specifying more intricate elements, the system will be protected against private key attacks.*

Remark 5.10. *It is obvious that, for the* $k(x) = \dfrac{p(x)}{q(x)}$ *function, if* $q(x) = 1$, *then the system will return its previous (classical) state. Therefore, if* $q(x) \neq 1$, *the system is more general.*

Remark 5.11. *If the* r *error function chosen arbitrarily in Equation 3 is chosen as* $r = p_1(x) * p_2(x) * q_2(x)$ *for* $f = \dfrac{p_1(x)}{q_1(x)}$ *and* $g = \dfrac{p_2(x)}{q_2(x)}$, *then the* $pr \circledast h$ *expression can be considered an equation in* $Q_p[X]$ *instead of* $K(x)$. *Thus, the operations are handled in* $Q_p[X]$, *and the algebraic structure of* $K(x)$ *will be used for only the* m *message function.*

5.10 GENERATION OF MESSAGE FUNCTIONS

A message function has the type $m = \dfrac{m_1(x)}{m_2(x)}$ for $m_1(x) = \sum_{i=0}^{n-1} \alpha_i x^i$ and $m_2(x) = \sum_{i=0}^{n-1} \beta_i x^i$. In the decoding phase, the message will be considered an $m = [\alpha_0, \alpha_1, ..., \alpha_{n-1}, \beta_0, \beta_1, ..., \beta_{n-1}]$ type element of Q^{2N}. In addition, if it is considered in the system suggestion given in the first section (writing a rational coefficient polynomial with the numerator and denominator consecutive), the message word comes to $4N$ lengths for classical NTRU systems, and therefore the proposed system is stronger against plain text attacks.

It can be easily shown that the set of rational functions is a vector space on Q_p. In this context, if a good base set is determined and the selected messages rewritten according to this base, the encryption is increased one more step.

Now, let's give a lemma for this purpose and suggest a new encryption.

Lemma 5.7. *For different* $q \in Q_p$ *numbers, the set* $\left(\dfrac{1}{x-q}\right)_{q \in Q_p}$ *is linearly independent.*

Proof. For $c \in Q_p$ scalars, consider the following equation:

$$\sum_{i=1}^{n-1} \frac{c_i}{x - q_i} = 0$$

To show that all q s are zero without disturbing generality, consider the equation

$$\frac{c_1}{x-q_1} = -\sum_{i=2}^{n-1} \frac{c_i}{x-q_i}$$

and show that $c_1 = 0$ in the equation.

Since x is variable and c_i is scalar, if $i \neq j$,

$$c_1\left(x-q_i\right) \neq c_1\left(x-q_j\right)$$

would be true. Obviously, this means that equality is only possible for the same q_i, and therefore all c_is are zero.

Now, a few lemmas will be given about how to write and send message functions on a different base instead of a standard base.

Lemma 5.8. *Let $q(x)$ be a reducible, and all of its roots are different from each other. For any $p(x)$, the function $\frac{p(x)}{q(x)}$ can be written in one type on the basis of $\left(\frac{1}{x-q}\right)$ and x^j.*

Proof. Let $q(x) = (x-q_0)(x-q_1)\ldots(x-q_{n-2})$.

Let's give the base order as:

$$\left\{x^0, x^1, \ldots, x^{n-1}, \frac{1}{x-q_1}, \frac{1}{x-q_2}, \ldots, \frac{1}{x-q_{n-1}}\right\}$$

So it will be possible to write any $m(x) = \frac{p(x)}{q(x)}$. A message function is of the type

$$\frac{p(x)}{q(x)} = \sum_{i=0}^{n-1} d_i x^i + c_i\left(\frac{1}{x-q_i}\right)$$

Therefore, the selection of d_i s and c_i s will give a new encoding message word.

Theorem 5.3. *For any message functions $m(x) = \frac{m_1(x)}{m_2(x)}$:*

Let μ be a operator such that

$$\mu(m(x)) = \sum d_i x^i + c_i\left(\frac{1}{x-q_i}\right)$$

That is, the μ operator rewrites the message function on a $\left\{x^j, \frac{1}{x-q_j}\right\}$ base. This type of μ operator can be added to the proposed cryptosystem.

Proof. After the classical decryption of NTRU algorithm process, the receiver arrives to the phase as:

$$e \equiv \mu(m(x))(modp)$$

and if the secret base set or μ operator is known, then the message function is available.

Now we will given a lemma for cases where the degree of the numerator polynomial is smaller than the polynomial in the denominator.

Lemma 5.9. *Let* $G(x)=\dfrac{m(x)}{S(x)}$ *be a rational function in* $Q_p[X]/(X^N-1)$ *with* $degm < degS$. *Also, let* S *be a separable polynomial. If the roots of* S *are:*

$$\{\alpha_1,\alpha_2,...,\alpha_S; S \le n-1\}$$

then there are $c_i \in Q_p$ *such that*

$$G(x)=\sum_{i=1}^{S}\frac{c_i}{(x-\alpha_i)}$$

will be realized.

Proof. Because $\dfrac{1}{x-\alpha_i}$ numbers are linearly independent for $\alpha_i \in Q_p$, if $i \ne j$, then $\dfrac{c_i}{x-\alpha_i} \ne \dfrac{c_j}{x-\alpha_j}$. It is clear that the denominator of this expression is max $S(x)$. Because Q_p is a field structure, it can be possible that $c_i\alpha_j = c_j\alpha_i$ whether j is different from i. Therefore, it can be seen that the $G(x)$ polynomial is the linear combination of $\dfrac{1}{x-\alpha_i}$ numbers.

Proposition 5.1. *In the proposed system, if the message function is written according to base* x^j *first and then resolved according to base* $\dfrac{1}{x-\alpha_j}$ *during the decryption process, the network will properly work under the situation in the previous lemma.*

Proof. $G(x)=\dfrac{m(x)}{S(x)}$ is represented by $g=(0,0,...,0,c_1,c_2,...,c_{n-1})$ on base $\left\{x^j,\dfrac{1}{x-\alpha_i}\right\}$. If this c_i's are chosen and send as $\sum c_i x^i$ in encryption phase, the message will be more securely hidden. At the end of the decryption phase, the receiver recombines this message for the $\dfrac{1}{x-\alpha_i}$ base and reaches $G(x)$ directly.

Remark 5.12. *For the* $C(x)=\sum c_i x^i$ *message polynomial, if the coefficients are chosen and produce a*

$$C(x)=\Sigma\frac{c_i}{x-\alpha_i}$$

function, because this function is written with a $\frac{1}{x-\alpha_i}$ *base, this is a base changing protocol. If this protocol is used, then the network will return the classical NTRU system. Because the receivers know that* $\Pi(x-\alpha_i)=S(x)$, *the polynomial is an error polynomial.*

5.11 AN EXAMPLE OF A DIGITAL SIGNATURE THAT CAN BE BUILT ON RATIONAL FUNCTIONS

A digital signature is used to verify the identity of the recipient in cryptosystems. Then this is an algebraic process that strengthens the system one more degree.

Now, let's first give a lemma about the structure that can be used as a signature in the system.

Lemma 5.10. *If a message function* $m(x)=\frac{f(x)}{g(x)}$ *has these two conditions:*

1. $degf < degg$
2. $g(x)$ *has only two factors; that is,* $g(x)=p(x)q(x)$ *and* $(p(x),q(x))=1$

then there is only one pair (s,t) *of specific polynomials such that* $m(x)=\frac{s(x)}{p(x)}+\frac{t(x)}{q(x)}$, $degs < degp$, *and* $degt < degq$ *can be written.*

Proof. If $(p,q)=1$, then there are $s,t\in Q_p[X]$ such that $ps+qt=1$ can be written. If both sides are divided by $p(x)q(x)$, then $\frac{s}{q}+\frac{t}{p}=\frac{1}{pq}$. Now, if both sides are multiplied by f

$$\frac{fs}{q}+\frac{ft}{p}=\frac{f}{g} \tag{5.4}$$

is reached. Now, if selections are made so that $degfs < degq$ and $degft < degp$, then $fs=s$ and $ft=t$. The proof ends with this pair of s and t if reached by choosing the appropriate parameters. On the contrary, if the numerator polynomials on the left of Equation 4 are of degrees greater then the denominator, then there are h,h,r,r in $Q_p[X]$ such that

$$\begin{aligned} fs &= qh+r,\ degr < degq \\ ft &= ph+r,\ degr < degp \end{aligned} \tag{5.5}$$

In the last case, Equation 4 would be

$$h+\frac{r}{q}+h+\frac{r}{p}=\frac{f}{g}$$

Because of the $degf < degg$ selections, the $h+h$ polynomial must be zero identically. And thus the process ends.

Now let's use the results of the previous lemma in the proposed cryptosystem.

Theorem 5.4. *Let* $m = \frac{f}{g}$ *be a message function with* $degf < degg$ *, and let* g *consist of only two coprime factors:* p *and* q *. If* e *is the encrypted function, then the* $\left(e, \frac{1}{p}, \frac{1}{q}\right)$ *signed message package will be properly decrypted.*

Proof. First, $m = \frac{f}{g}$ is obtained from e encrypted functions. Because g has only two linearly independent factors, the m message function is uniquely represented for this base. That is, if $pq = g$ and $(p,q) = 1$, then the $\frac{r}{p} + \frac{r}{q} = m$ expression is possible for a unique r and r in $Q_p[x]$. Therefore, the (m,r,r) package is the final process for this network.

5.12 CONCLUSION AND RECOMMENDATIONS

In this study, the NTRU cryptosystem is mainly discussed in a larger algebraic structure. With some new results, the system has been made more protected against plaintext attacks and private key attacks. In addition, a new digital signature or authentication method has been given. By attacking or cryptoanalyzing the outputs of this study, different implementation methods can be given.

REFERENCES

[1] J. Hoffstein, J. Pipher and J. H. Silverman, A ring-based public key cryptosystem, *ANTS 1998*, LNCS, Springer, Vol. 1423 (1998), 267–288.

[2] J. Hoffstein and J. Silverman, Optimizations for NTRU, in: *Public-Key Cryptography and Computational Number Theory*, Degruyter (1971).

[3] J. Hoffstein, J. Pipher, J. H. Silverman and W. Whyte, $NTRU_{SIGN}$: Digital signatures using the NTRU lattice, *LNCS*, Springer, Vol. 2612 (2003).

[4] J. H. Silverman, Invisibility in truncated polynomial rings, *Technical Report*, NTRU Cryptosystems (2003). Available at www.ntru.com.

[5] N. A. Graham, P. Q. Nguyen, J. Proos, J. H. Silverman, A. Singer and W. Whyte, The impact of decryption failures on the security of NTRU encryption, *Crypto'03, LNCS*, Springer-Verlag, Vol. 2729 (2003), 226–246.

[6] J. H. Silverman and W. Whyte, Estimating decryption failure probabilities for NTRU encrypt, *Technical Report 18* (2005). Available at www.ntru.com.

[7] J. Hoffstein, J. Pipher, J. H. Silverman, W. Whyte and Z. Zhang, Choosing parameters for NTRU encrypt. *NTRU Challenge* (2015). Available at www.ntru.com/ntru-challenge.

[8] N. Abdurrahmane, *The Mathematics of the NTRU Public Key Cryptosystem*, Mathematical Concepts IGI Global (2015).

[9] D. Coppersmith and A. Shamir, Lattice attacks on NTRU, *Euro-crypto'97*, Lecture Notes in Computer Science, Vol. 1233, Springer, Berlin (1997).

[10] J. Hoffstein, J. H. Silverman and W. Whyte, Estimating breaking times for NTRU lattices, *Technical Report 12* (2003). Available at www.ntru.com.

[11] N. Graham, J. H. Silverman and W. Whyte, A meet in the middle attack on a NTRU private key, *NTRU Technical Report 04* (2003). Available at www.ntru.com.

[12] N. Graham, J. H. Silverman, A. Singer and W. Whyte, NAEP: Provable security in the presence of decryption failures, *Cryptology ePrint Archive, Report 2003/172* (2003).

[13] W. Whyte, N. Graham and J. H. Silverman, *Choosing Parameter Sets for NTRU Encrypt with NAEP and SVES-3*, Topics in Cryptology CT-RSA (2005).

[14] T. Meskanen and A. Renuall, A wrap error attack against NTRU, *University of Turku Technical Report TUCS 507* (2003).

[15] J. Proos, *Imperfect Decryption and an Attack on the NTRU* (2003). Available at http://eprint.iacr.org.

[16] J. Hong, J. W. Han and D. Han, *Chosen-Ciphertext Attacks on Optimized NTRU* (2002). Available at http://eprint.iacr.org.

[17] P. Nguyen and D. Pointcheval, Analysis and improvements of NTRU paddings, *Crypto 2002*, Springer-Verlag (2002).

[18] C. Gentry and M. Szydlo, Cryptoanalysis of the revised NTRU signature scheme, *Eurocrypt'02*, Vol. 2332, Springer-Verlag (2002).

[19] A. Çağman, K. Polat and S. Taş, A key agreement protocol based on group actions, *Numerical Methods for Partial Differential Equations*, 37(2), 1112–1119 (2021).

[20] A. Çağman, A key agreement protocol involving the Taylor polynomials of differentiable functions, *Fundamentals of Contemporary Mathematical Sciences*, 2(2), 121–126 (2021).

6 Metric Space and Fixed-Point Results in $G-$ Metric Spaces

Talat Nazir and Mujahid Abbas

6.1 INTRODUCTION TO G – METRIC SPACE

Fixed point theory is a useful and powerful tool for solving several kind of problems arising in nonlinear phenomena. Dhage [1] defined the notion of $D-$ metric space in an attempt to obtain analogous results to those for metric spaces, but in a more general setting. In the listed papers (including [2–4]), Dhage presented topological structures in such spaces together with several fixed point results. These works also show a substantial number of results by other authors. Unfortunately, most of the proposed results of Dhage [1] related to the basic properties of topology are not correct.

To overcome fundamental flaws in Dhage's theory of generalized metric spaces [1], Mustafa and Sim [5] gave a more appropriate generalization of metric, that of $G-$ metric space. Afterwards, Mustafa and Sim [6] generalized the structure of space by employing a structure of the $G-$ metric. Several extended new results related to fixed points of self maps involving general contractive constraints are established in Mustafa et al. [5, 7–11]. Saadati et al. [12] employed the concept of a partially ordered $G-$ metric and presented some fixed-point results. Further interesting results in this direction are obtained in [13–20].

We begin with the following definition given by Mustafa and Sims [5] of the $G-$ metric.

Definition 6.1. *For a nonempty set X, if a map $\mathcal{G}: X \times X \times X \to [0,\infty)$ satisfies:*

$\mathcal{G}_1$) $\mathcal{G}(s,r,x)=0$ *if and only if* $s=r=x$,

$\mathcal{G}_2$) $0<\mathcal{G}(s,r,x)$ *for all* $s,r \in X$, *having* $r \neq x$,

$\mathcal{G}_3$) $\mathcal{G}(s,s,r) \leq \mathcal{G}(s,x,r)$ *for all* $s,r,x \in X$, *having* $x \neq r$,

$\mathcal{G}_4$) $\mathcal{G}(r,x,s)=\mathcal{G}(s,x,r)=\mathcal{G}(s,r,x)=\ldots$ *(symmetric distance property), and*

$\mathcal{G}_5$) $\mathcal{G}(a,s,r) \leq \mathcal{G}(a,x,x)+\mathcal{G}(x,s,r)$ *for all* $s,a,x,r \in X$.

Then map $\mathcal{G}$ is known as a $G-$ metric on X and a pair $(X,\mathcal{G})$ is known as a s $G-$ metric space.

DOI: 10.1201/9781003322856-6

Definition 6.2. *In a $G-$ metric space $(X,\mathcal{G})$, if we have $\mathcal{G}(r,s,s)=\mathcal{G}(r,s,s)$ for all $r,s\in X$, then we say that $\mathcal{G}$ is a symmetric $G-$ metric on X.*

Proposition 6.1. *In a $G-$ metric space $(X,\mathcal{G})$, we have a metric $d_{\mathcal{G}}$ on X as*

$$d_{\mathcal{G}}(r,u)=\mathcal{G}(r,u,u)+\mathcal{G}(r,r,u),\forall r,u\in X \tag{6.1}$$

In the case where we have a symmetric $G-$ metric space $(X,\mathcal{G})$, then $d_{\mathcal{G}}$ becomes

$$d_{\mathcal{G}}(r,u)=2\mathcal{G}(r,u,u),\forall r,u\in X \tag{6.2}$$

Otherwise, the following inequality holds

$$\frac{3}{2}\mathcal{G}(r,u,u)\le d_{\mathcal{G}}(r,u)\le 3\mathcal{G}(r,u,u),\forall r,u\in X \tag{6.3}$$

Example 6.1. *If $X=[\alpha,\beta]$ for some real $\alpha,\beta\in[0,\infty)$ with $\alpha,<\beta$, then $\mathcal{G}(r,u,s)=$ $max\{|r-u|,|u-s|,|s-r|\}$ is a $\mathcal{G}-$metric on X.*

Example 6.2. *If $X=\{1,2\}$, then $\mathcal{G}:X\times X\times X\to[0,\infty)$ obtained by*

(r,u,s)	$\mathcal{G}(r,u,s)$
$(1,1,1),(2,2,2)$	0
$(1,1,2),(1,2,1),(2,1,1)$	0.5
$(1,2,2),(2,1,2),(2,2,1)$	1

is a non-symmetric $G-$ metric on X as $\mathcal{G}(r,s,s)\ne\mathcal{G}(r,r,s)$ for distinct r,s in X.

Example 6.3. *If for some natural numbers α,β,γ, we have $X=\{\alpha,\beta,\gamma\}$, the $\mathcal{G}:X\times X\times X\to[0,\infty)$ given as*

(r,u,s)	$\mathcal{G}(r,u,s)$
$(\alpha,\alpha,\alpha),(\beta,\beta,\beta),(\gamma,\gamma,\gamma),$	0
$(\alpha,\beta,\beta),(\beta,\alpha,\beta),(\beta,\beta,\alpha),$	1
$(\alpha,\alpha,\beta),(\alpha,\beta,\alpha),(\beta,\alpha,\alpha),$ $(\beta,\gamma,\gamma),(\gamma,\beta,\gamma),(\gamma,\gamma,\beta),$	2
$(\alpha,\alpha,\gamma),(\alpha,\gamma,\alpha),(\gamma,\alpha,\alpha),$ $(\alpha,\gamma,\gamma),(\gamma,\alpha,\gamma),(\gamma,\gamma,\alpha),$	3
$(\beta,\beta,\gamma),(\beta,\gamma,\beta),(\gamma,\beta,\beta),$ $(\alpha,\beta,\gamma),(\alpha,\gamma,\beta),(\beta,\alpha,\gamma),$ $(\beta,\gamma,\alpha),(\gamma,\alpha,\beta),(\gamma,\beta,\alpha),$	4

is a non-symmetric $G-$ metric, since $\mathcal{G}(\alpha,\alpha,\gamma)\ne\mathcal{G}(\alpha,\gamma,\gamma)$.

Proposition 6.2. *For a given $G-$ metric space $(X,\mathcal{G})$,*

$$(E_1)\ \mathcal{G}_1(r,u,s)=min\{\kappa,\mathcal{G}(r,u,s)\} \text{ for some } \kappa>0$$

$$(E_2)\ \mathcal{G}_2(r,u,s)=\frac{\mathcal{G}(r,u,s)}{1+\mathcal{G}(r,u,s)}$$

are also $G-$ metrics on X. Further, if s $X=\cup_{i=1}^{n}A_i$ is any partition of X and for $\kappa>0$, then

$$(E_3)\,\mathcal{G}_3(\boldsymbol{r},\boldsymbol{u},\boldsymbol{s})=\begin{cases}\mathcal{G}(\boldsymbol{r},\boldsymbol{u},\boldsymbol{s}), \text{ if for some } \boldsymbol{i} \text{ we have } \boldsymbol{r},\boldsymbol{u},\boldsymbol{s}\in A_i\\ \kappa+\mathcal{G}(\boldsymbol{r},\boldsymbol{u},\boldsymbol{s}), \text{ otherwise}\end{cases}$$

is a $G-$ metric on X.

We can derive the following useful properties from a given $G-$ metric on X.

Proposition 6.3. *For a given $G-$ metric space $(X,\mathcal{G})$, for any $r,u,s,y,x\in X$, we have the following.*

(1) *if* $\mathcal{G}(r,u,s)=0$, *then* $r=s=u$,
(2) $\mathcal{G}(r,y,s)\le\mathcal{G}(r,r,y)+\mathcal{G}(y,s,y)$,
(3) $\mathcal{G}(r,r,u)\le 2\mathcal{G}(u,r,u)$,
(4) $\mathcal{G}(r,u,s)\le\mathcal{G}(r,x,s)+\mathcal{G}(x,u,s)$,
(5) $\mathcal{G}(r,u,s)\le\frac{2}{3}\left[\mathcal{G}(r,u,y)+\mathcal{G}(r,y,s)+\mathcal{G}(y,u,s)\right]$,
(6) $\mathcal{G}(r,u,s)\le\mathcal{G}(r,y,y)+\mathcal{G}(y,u,y)+\mathcal{G}(x,x,s)$,
(7) $\left|\mathcal{G}(r,u,s)-\mathcal{G}(r,u,y)\right|\le\max\{\mathcal{G}(y,s,s),\mathcal{G}(s,y,y)\}$,
(8) $\left|\mathcal{G}(r,u,s)-\mathcal{G}(r,u,x)\right|\le\mathcal{G}(r,x,s)$,
(9) $\left|\mathcal{G}(r,u,s)-\mathcal{G}(u,s,s)\right|\le\max\{\mathcal{G}(r,s,s),\mathcal{G}(s,r,r)\}$,
(10) $\left|\mathcal{G}(r,y,y)-\mathcal{G}(y,r,r)\right|\le\max\{\mathcal{G}(y,r,r),\mathcal{G}(r,y,y)\}$.

Definition 6.3. *In a $G-$ metric space $(X,\mathcal{G})$, for $u_0\in X$, $\kappa>0$, a $G-$ ball with center u_0 along with radius κ is*

$$\mathcal{B}_{\mathcal{G}}(u_0,\kappa)=\{r\in X\mid\mathcal{G}(u_0,r,r)<\kappa\}$$

Proposition 6.4. *In a $G-$ metric space $(X,\mathcal{G})$ with $u_0\in X$, $\kappa>0$, we have*

(1) *if* $\mathcal{G}(u_0,r,u)<\kappa$, *then* $r,u\in\mathcal{B}_{\mathcal{G}}(u_0,\kappa)$;
(2) *if* $y\in B_{\mathcal{G}}(u_0,\kappa)$, *then we can find a* $\delta>0$ *such that* $\mathcal{B}_{\mathcal{G}}(y,\delta)\subseteq\mathcal{B}_{\mathcal{G}}(u_0,\kappa)$.

Proof. *(1) It follow from ($\mathcal{G}_3$) that $\mathcal{G}(u_0,r,r)\le\mathcal{G}(u_0,r,u)<\kappa$ and $\mathcal{G}(u_0,u,u)\le\mathcal{G}(u_0,r,u)<\kappa$, which implies that $r,u\in\mathcal{B}_{\mathcal{G}}(u_0,r)$.*

(2) follows from ($\mathcal{G}_5$) with $\delta = \kappa - G(u_0, y, y)$. It follows from (2) of the previous proposition that the family of all $G-$ balls

$$\beta = \mathcal{B}_{\mathcal{G}}(u_0, \kappa) = \{r \in X : G(u_0, r, r) < \kappa\}$$

is the base of a topology τ_G on X.

Proposition 6.5. *In a $G-$ metric space $(X, \mathcal{G})$, for any $u_0 \in X$, $\kappa > 0$, we have*

$$\mathcal{B}_{\mathcal{G}}\left(u_0, \frac{1}{3}\kappa\right) \subseteq \mathcal{B}_{d_{\mathcal{G}}}(u_0, \kappa) \subseteq \mathcal{B}_{\mathcal{G}}(u_0, \kappa) \tag{6.4}$$

Consequently, the $G-$ metric topology $\tau(G)$ on X is similar to the metric topology of $d_{\mathcal{G}}$.

6.2 CONVERGENCE AND CONTINUITY IN G – METRIC SPACES

Definition 6.4. *In a $G-$ metric space $(X, \mathcal{G})$, a sequence $\{r_n\}$ in X is $G-$ convergent to r if it converges to r in $G-$ metric topology, $\tau(G)$.*

Proposition 6.6. *In a $G-$ metric space $(X, \mathcal{G})$, the following statements are equivalent:*

(1) The sequence $\{r_n\}$ is $G-$ convergent to r.
(2) $\mathcal{G}(r_n, r_n, r) \to 0$ as $n \to \infty$.
(3) $\mathcal{G}(r_n, r, r) \to 0$ as $n \to \infty$.
(4) $\mathcal{G}(r_n, r_m, r) \to 0$ as $n, m \to \infty$.

Definition 6.5. *Let $(X, \mathcal{G})$, $(X', \mathcal{G}')$ be $G-$ metric spaces. A self map $f : X \to X$ is $G-$ continuous at element u_0 in X whenever $f^{-1}\left(B_{G'}(f(u_0), r)\right) \in \tau(G)$, for all $r > 0$. We say f is $G-$ continuous when f is $G-$ continuous at all points belonging to X, that is, continuous as a function from X with the $\tau(G)-$ topology to X' with $\tau(G')-$ topology.*

Since $G-$ metric topologies are metric topologies, we have:

Proposition 6.7. *For two $G-$ metric spaces $(X, \mathcal{G})$ and (X', G'), a self map $f : X \to X$ is $G-$ continuous on an element $r \in X$ if and only if f is $G-$ sequentially continuous at u; that is, the case where $\{r_n\}$ is $G-$ convergent to r implies $\{f(r_n)\}$ is $G-$ convergent to $f(r)$.*

Proposition 6.8. *In a $G-$ metric space $(X, \mathcal{G})$, if sequences $\{r_l\}$, $\{u_n\}$, and $\{s_m\}$ are sequences in X such that $r_l \to r$, $u_n \to u$, and $s_m \to s$, then $\mathcal{G}(r_l, u_n, s_m) \to \mathcal{G}(r, u, s)$ as $l, n, m \to \infty$.*

Proof. *As $\{r_l\}$, $\{u_n\}$, $\{s_m\}$ converges to r, u, s, respectively, so, by ($\mathcal{G}_5$), it follows that*

$$\mathcal{G}(r,u,s) \leq \mathcal{G}(u,u_n,u_n)+\mathcal{G}(u_n,r,z)$$
$$\mathcal{G}(r,u_n,s) \leq \mathcal{G}(r,r_l,r_l)+\mathcal{G}(r_l,u_n,s)$$
$$\text{and}\,\mathcal{G}(r_l,u_n,s) \leq \mathcal{G}(r_l,u_n,s_m)+\mathcal{G}(s_m,s_m,s)$$

Therefore

$$\mathcal{G}(r,u,s)-\mathcal{G}(r_l,u_n,s_m)$$
$$\leq \mathcal{G}(r_l,r_l,r)+\mathcal{G}(u_n,u_n,u)+\mathcal{G}(s_m,s_m,s)$$

Similarly,

$$\mathcal{G}(r_l,u_n,s_m)-\mathcal{G}(r,u,s)$$
$$\leq \mathcal{G}(r_l,r,r)+\mathcal{G}(u_n,u,u)+\mathcal{G}(s_m,s,s)$$

Combining these using Proposition 1.3, we obtain

$$\left|\mathcal{G}(r_l,u_n,s_m)-\mathcal{G}(r,u,s)\right|$$
$$\leq 2\left[\mathcal{G}(r_l,r_l,r)+\mathcal{G}(u_n,u_n,u)+\mathcal{G}(s_m,s_m,s)\right]$$

and on taking the limit as $l,n,m \to \infty$, we get the conclusion.

Proposition 6.9. In a $G-$ metric space $(X,\mathcal{G})$, if the sequence $\{r_n\}$ converges to r, and $\{r_n\}$ converges to s, for some $r,s \in X$. Then $r = s$.

Proof. By $(\mathcal{G}_5)$, we have

$$\mathcal{G}(r,r,s) \leq \mathcal{G}(r_m,r_n,s)+\mathcal{G}(r_m,r_n,r)$$

and

$$\mathcal{G}(r_n,r,s) \leq \mathcal{G}(r_m,r_m,s)+\mathcal{G}(r_m,r_n,r)$$

So

$$\mathcal{G}(r,r,s) \leq \mathcal{G}(r_n,r_n,r)+\mathcal{G}(r_m,r_n,s)+\mathcal{G}(r_m,r_n,r)$$

By limiting as $n,m \to \infty$ implies $\mathcal{G}(r,r,s) = 0$. Therefore, $r = s$.

Definition 6.6. In a $G-$ metric space $(X,\mathcal{G})$, a sequence $\{r_n\}$ is

(1) a $G-$ Cauchy sequence if for $\varepsilon > 0$, there is $k_0 \in \mathbb{N}$; that is, for all $l,m,n \geq k_0$, $\mathcal{G}(r_l,r_m,r_n) < \varepsilon$;
(2) a $G-$ convergent sequence if for $\varepsilon > 0$, there is $r \in X$ and $k_0 \in \mathbb{N}$; that is, for all $l,n \geq k_0$, $\mathcal{G}(r,r_l,r_n) < \varepsilon$.

In the case where every $G-$ Cauchy sequence in X is convergent in X, then such a $G-$ metric space $(X,\mathcal{G})$ is known as complete. Note then if $\mathcal{G}(r_l,r_n,r)\to 0$ as $l,n\to\infty$, then we say that $\{r_n\}$ converges to r in $(X,\mathcal{G})$.

Proposition 6.10. In a $G-$ metric space $(X,\mathcal{G})$, the following are equivalent:

(1) If $\varepsilon>0$, there is $k_0\in\mathbb{N}$; that is, for all $l,n\geq k_0$, $\mathcal{G}(r_l,r_n,r_n)<\varepsilon$; that is, $\mathcal{G}(r_l,r_n,r)\to 0$ as $l,n\to\infty$.
(2) Sequence $\{r_n\}$ is $G-$ Cauchy in $(X,\mathcal{G})$.

Corollary 6.1. *In a $G-$ metric space $(X,\mathcal{G})$, if a $G-$ Cauchy sequence has a $G-$ convergent subsequence, then sequence itself is $G-$ convergent.*

Corollary 6.2. *If $Z\neq\phi$ is a set contained in $G-$ complete metric space $(X,\mathcal{G})$, then $\left(Y,G|_Y\right)$ is complete whenever Y is $G-$ closed in $(X,\mathcal{G})$.*

Definition 6.7. *An element ω in set X is known as a fixed point of a map $f:X\to X$ when $\omega=f\omega$. We denoted $F(f)$ as a collection of all fixed points of f.*

As in [21], we develop the following definition in $G-$ metric space.

Definition 6.8. *In a $G-$ metric space $(X,\mathcal{G})$, a map $T:X\to X$ is called:*

(1) *continuous at u in X when any $\{s_n\}$ in X $G-$ convergent to s implies $\{Ts_n\}$ $G-$ convergent to Ts,*
(2) *sequentially convergent when any sequence $\{u_n\}$ in X that has $\{Ts_n\}$ is $G-$ convergent in X implies $\{s_n\}$ is $G-$ convergent in X,*
(3) *subsequentially convergent when a sequence $\{r_n\}$ in X having $\{Tr_n\}$ is $G-$ convergent in X implies $\{r_n\}$ has a $G-$ convergent subsequence in X.*

6.3 FIXED-POINT RESULTS IN $G-$ METRIC SPACE

For $G-$ metric space $(X,\mathcal{G})$, consider a self map $f:X\to X$. A collection $O_f(s)=\{s,fs,f^2s,\ldots\}$ is known as an orbit of s [22]. f is called orbitally continuous if $u=\lim_i f^{m_i}x$ implies $fu=\lim_i f\left(f^{m_i}x\right)$. Notice that every continuous self-mapping is orbitally continuous but not conversely [23].

Now, we consider the following theorem.

Theorem 6.1. In a $G-$ metric space $(X,\mathcal{G})$, if there is $u\in X$ and $\lambda\in[0,1)$ with $\overline{O(u)}$ complete and

$$\mathcal{G}(fr,fs,fs)\leq\lambda\mathcal{G}(r,s,x) \tag{6.5}$$

for each r, $s=x=fr\in O(u)$. Then $\{f^n r\}$ converges to some point $p\in X$, and, if for $l,n\in\mathbb{N},l>n$,

$$\mathcal{G}(r_n, r_l, r_l) \leq \frac{\lambda^n}{1-\lambda} \mathcal{G}(fu, fu, u)$$

for $n \geq 1$. Further, if if (1.5) holds for all $x \in \overline{O(u)}$ or f is orbitally continuous at p, then p is a fixed point of f.

Proof. In the case of a symmetric $G-$ metric, we have

$$d_{\mathcal{G}}(r,s) = 2\mathcal{G}(r,s,s)$$

(6.5) becomes

$$d_{\mathcal{G}}(fr, fs) \leq \lambda d_{\mathcal{G}}(r,s)$$

for all $s \in O(u)$, and we get the proof from Theorem 6.2 of [24]. For a non-symmetric $G-$ metric, consider $r_n := f^n u$; (6.5) implies

$$\mathcal{G}(r_{n+1}, r_{n+2}, r_{n+2}) \leq \lambda \mathcal{G}(r_n, r_{n+1}, r_{n+1}) \leq \cdots \leq \lambda^n \mathcal{G}(u, fu, fu)$$

For all $m, n \in \mathbb{N}, m > n$, from property $(\mathcal{G}_5)$, it follows that

$$\begin{aligned} \mathcal{G}(r_n, r_m, r_m) &\leq \mathcal{G}(r_n, r_{n+1}, r_{n+1}) + \mathcal{G}(r_{n+1}, r_{n+2}, r_{n+2} + \\ &\cdots + \mathcal{G}(r_{m-1}, r_m, r_m) \\ &\leq \left(\lambda^n + \lambda^{n+1} + \cdots + \lambda^{m-1}\right) \mathcal{G}(u, fu, fu) \\ &\leq \frac{\lambda^n}{1-\lambda} \mathcal{G}(u, fu, fu) \end{aligned}$$

Therefore, the limit of sequence $\{r_n\}$ exists. We call the limit p. If f is orbitally continuous at $r = p$, then $\lim_{n\to\infty} f^n u = p$ gives $\lim_{n\to\infty}\left(f^{n+1}u\right) = fp$, and p is a fixed point of f. If (6.5) holds for $r, y = z \subset O(u)$, then the inequality

$$\mathcal{G}\left(fp, f^{n+1}u, f^{n+1}u\right) \leq \lambda \mathcal{G}\left(p, f^n u, f^n u\right)$$

implies that $p = fp$.

Theorems 6.1, 6.2, and 6.5 and Corollaries 1 and 2 of [7] are special cases of Theorem 6.1 in the following sense.

Consider Theorem 6.1 of [7] with the contractive condition

$$\begin{aligned} \mathcal{G}(fr, fs, fy) \leq\ & a\mathcal{G}(r,s,y) + b\mathcal{G}(r, fr, fr) + c\mathcal{G}(y, fy, fy) \\ & + d\mathcal{G}(s, fs, fs) \end{aligned} \tag{6.6}$$

Notice that if f has a fixed point, then (6.6) implies uniqueness. If G is symmetric, setting $s = fr$ yields

$$d_{\mathcal{G}}\left(fr, f^2 r\right) \leq k d_{\mathcal{G}}\left(r, fr\right) \tag{6.7}$$

where $k = \dfrac{b+c+d}{2-b-c-d} < 1$. Thus, if $r_n = f^n r_0$ for any $r_0 \in X$, then (6.7) is true for all $r \in \overline{O(r_0)}$, and, from Theorem 6.2 of [24], f has a fixed point p. Similarly, the case where G is non-symmetric implies one obtains, as in [7],

$$\mathcal{G}\left(r_n, r_{n+1}, r_{n+1}\right) \leq q\mathcal{G}\left(r_{n-1}, r_n, r_n\right)$$

for some $0 \leq q < 1$, and a fixed point of f follows from Theorem 6.1.

Example 6.4. *Consider $X = [0,1]$ with $\mathcal{G}(r,s,r) = max\{|r-s|, |s-r|, |r-r|\}$ given a $G-$ metric on X. A map $f: X \to X$ defined as*

$$f(r) = \begin{cases} \dfrac{r}{2} & \text{if } r \in \left[0, \dfrac{3}{4}\right) \\ \dfrac{1}{2} & \text{if } r \in \left[\dfrac{3}{4}, 1\right] \end{cases}$$

Take $u = \frac{1}{2} \in X$; then $\overline{O(u)} = \left\{0, \frac{1}{2}, \frac{1}{4}, \frac{1}{8}, \ldots\right\}$, and we can verify that $\mathcal{G}(fr, fs, fs) \leq \lambda\mathcal{G}(r,s,r)$ for each r, $s = r = fr \in O(u)$, where $\lambda = \frac{1}{2}$.

Obviously $\{f^n x\}$ converges to 0, and at $0 \in X$, f is orbitally continuous. Moreover, $0 \in F(f)$.

Note that if we define $f: X \to X$ as $u = fu$, then for any arbitrary fixed u in X, $\overline{O(u)} = \{u\}$. Moreover, $\mathcal{G}(fr, fs, fs) \leq \lambda\mathcal{G}(r,s,x)$ for each r, $x = s = fr \in O(u)$ is satisfied, $\{f^n u\}$ converges to $u \in X$, and f is orbitally continuous at u, which is a fixed point of f.

Note that f does not satisfy the contraction condition in $G-$ metric space; that is,

$$\mathcal{G}\left(f\left(\frac{2}{3}\right), f\left(\frac{3}{4}\right), f\left(\frac{3}{4}\right)\right) = \left|\frac{1}{3} - \frac{1}{2}\right| = \frac{1}{6} \not\leq \frac{\lambda}{12} = \lambda\mathcal{G}\left(\frac{2}{3}, \frac{3}{4}, \frac{3}{4}\right)$$

Also, for $x = y$, $z = 0$, f does not satisfy conditions (6.1) or (6.2) of Theorem 6.1 of [7].

6.4 FIXED POINT OF SELF-MAPS IN PARTIALLY ORDERED G – METRIC SPACES

Initially, the fixed point results under ordered metric spaces are obtained Ran and Reurings [25], Then other extended results are established by Nieto and Lopez [26].

Definition 6.9. *For a nonempty set X, $(X,\preceq,\mathcal{G})$ is known as ordered $G-$ metric space if*

(1) $(X,\mathcal{G})$ is $G-$ metric space, and

(2) $(X,\preceq)$ is a partial ordered set.

Definition 6.10. *In a partial ordered set $(X,\preceq)$, elements r and s of X are known as comparable if either $r\preceq s$ or $s\preceq r$.*

A subset $\mathcal{U}$ of $(X,\preceq)$ is called well ordered (totally ordered) if every element r and s of $\mathcal{U}$ is comparable.

We obtain fixed-point theorems of generalized $(\varphi,\psi)-$ weak contractive maps in $(X,\preceq,\mathcal{G})$, where φ and ψ are from the following two class mappings:

$\Phi=\{\varphi\mid\varphi:[0,\infty)\to[0,\infty)$ is nondecreasing and lower semi-continuous with $\varphi(r)>0$ for all $r>0$, $\varphi(0)=0\}$.

$\Psi=\{\psi\mid\psi:[0,\infty)\to[0,\infty)$ is nondecreasing and continuous with $\psi(r)=0$ if and only if $r=0\}$.

Theorem 6.2. *In a complete $G-$ metric space $(X,\preceq,\mathcal{G})$, if a self map f on X is nondecreasing and for all comparable $r,s,u\in X$,*

$$\psi(\mathcal{G}(fr,fs,fv))\le\psi(M(r,s,v))-\varphi(M(r,s,v)) \tag{6.8}$$

holds, where $\varphi\in\Phi$, $\psi\in\Psi$ and

$$\begin{aligned} M(r,s,v) &= a_1\mathcal{G}(r,s,v)+a_2\mathcal{G}(r,fr,fr)+a_3\mathcal{G}(s,fs,fs)+a_4\mathcal{G}(v,fv,fv)\\ &\quad +a_5\left[\mathcal{G}(r,fs,fs)+\mathcal{G}(s,fv,fv)+\mathcal{G}(u,fr,fr)\right]\end{aligned}$$

for $0<a_i$ for all $i=1,\ldots,5$ with $\sum_{i=1}^{5}a_i<1$. If there is an $r_0\in X$ with $fr_0\succeq r_0$ and either f is G-continuous or a nondecreasing sequence $\{r_n\}$ with $r_n\to z$ in X implies $z_n\preceq z$ for all $n\in\mathbb{N}$, then f has at most one fixed point.

Proof. *By a nondecreasing map, f implies*

$$r_1=fr_0\preceq f^2r_0\preceq\cdots\preceq f^nr_0\preceq f^{n+1}r_0\preceq\cdots$$

Define a sequence $\{r_n\}$ in X with $r_n=f^nr_0$, and so $r_{n+1}=fr_n$ for $n\ge0$. If $r_l=fr_l$ at some l, then r_l becomes a fixed point. Let us take $r_n\ne fr_n$ for all $n\in\mathbb{N}$. As $r_n\preceq r_{n+1}$ for every n, from (6.8),

$$\begin{aligned}&\psi\left(\mathcal{G}\left(r_{n+3},r_{n+2},r_{n+1}\right)\right)\\&=\psi\left(\mathcal{G}\left(fr_{n+2},fr_{n+1},fr_{n}\right)\right)\\&\le\psi\left(M\left(r_{n},r_{n+1},r_{n+2}\right)\right)-\varphi\left(M\left(r_{n},r_{n+1},r_{n+2}\right)\right)\end{aligned}\tag{6.9}$$

where

$$\begin{aligned}&M\left(r_{n},r_{n+1},r_{n+2}\right)\\&=a_1\mathcal{G}\left(r_{n+2},r_{n+1},r_{n}\right)+a_2\mathcal{G}\left(r_{n},fr_{n},fr_{n}\right)\\&\quad+a_3\mathcal{G}\left(r_{n+1},fr_{n+1},fr_{n+1}\right)+a_4\mathcal{G}\left(r_{n+2},fr_{n+2},fr_{n+2}\right)\\&\quad+a_5\left[\mathcal{G}\left(r_{n},fr_{n+1},fr_{n+1}\right)+\mathcal{G}\left(r_{n+1},fr_{n+2},fr_{n+2}\right)+\mathcal{G}\left(r_{n+2},fr_{n},fr_{n}\right)\right]\\&=a_1\mathcal{G}\left(r_{n},r_{n+1},r_{n+2}\right)+a_2\mathcal{G}\left(r_{n},fr_{n},fr_{n}\right)+a_3\mathcal{G}\left(r_{n+1},fr_{n+1},fr_{n+1}\right)\\&\quad+a_4\mathcal{G}\left(r_{n+2},fr_{n+2},fr_{n+2}\right)\\&=a_3\mathcal{G}\left(r_{n+1},r_{n+2},r_{n+2}\right)+a_4\mathcal{G}\left(r_{n},r_{n+2},r_{n+2}\right)\\&=\left(a_3+a_4\right)\mathcal{G}\left(r_{n+2},r_{n+2},r_{n+1}\right)\end{aligned}$$

So

$$\begin{aligned}&\psi\left(\mathcal{G}\left(r_{n+2},r_{n+2},r_{n+1}\right)\right)\\&\le\psi\left(\left(a_3+a_4\right)\mathcal{G}\left(r_{n+1},r_{n+2},r_{n+2}\right)\right)-\varphi\left(\left(a_3+a_4\right)\mathcal{G}\left(r_{n+2},r_{n+2},r_{n+1}\right)\right)\end{aligned}$$

implies $\varphi\left(\left(a_3+a_4\right)\mathcal{G}\left(r_{n+2},r_{n+2},r_{n+1}\right)\right)=0$ and so $r_{n+2}=r_{n+1}$. Thus r_{n+1} is a fixed point of f.

Going back to the assumption $\mathcal{G}\left(r_{n},r_{n+1},r_{n+1}\right)>0$ for $n\ge 0$, consider

$$\begin{aligned}&\psi\left(\mathcal{G}\left(r_{n+2},r_{n+2},r_{n+1}\right)\right)\\&=\psi\left(\mathcal{G}\left(fr_{n},fr_{n+1},fr_{n+1}\right)\right)\\&\le\psi\left(M\left(r_{n+1},r_{n+1},r_{n}\right)\right)-\varphi\left(M\left(r_{n},r_{n+1},r_{n+1}\right)\right)\end{aligned}\tag{6.10}$$

where

$$\begin{aligned}&M\left(r_{n+1},r_{n+1},r_{n}\right)\\&=a_1\mathcal{G}\left(r_{n+1},r_{n+1},r_{n}\right)+a_2\mathcal{G}\left(r_{n},fr_{n},fr_{n}\right)+a_3\mathcal{G}\left(r_{n+1},fr_{n+1},fr_{n+1}\right)\\&\quad+a_4\left[\mathcal{G}\left(r_{n},fr_{n+1},fr_{n+1}\right)+\mathcal{G}\left(r_{n+1},fr_{n},fr_{n}\right)\right]\end{aligned}$$

$$\begin{aligned}
&= a_1\mathcal{G}(r_{n+1},r_{n+1},r_n)+a_2\mathcal{G}(r_n,r_{n+1},r_{n+1})+a_3\mathcal{G}(r_{n+2},r_{n+2},r_{n+1})\\
&\quad +a_4\left[\mathcal{G}(r_n,r_{n+2},r_{n+2})+\mathcal{G}(r_{n+1},r_{n+1},r_{n+1})\right]\\
&\le (a_1+a_2)\mathcal{G}(r_n,r_{n+1},r_{n+1})+a_3\mathcal{G}(r_{n+1},r_{n+2},r_{n+2})\\
&\quad +a_4\left[\mathcal{G}(r_n,r_{n+1},r_{n+1})+\mathcal{G}(r_{n+1},r_{n+2},r_{n+2})\right]\\
&= (a_1+a_2+a_4)\mathcal{G}(r_n,r_{n+1},r_{n+1})+(a_3+a_4)\mathcal{G}(r_{n+2},r_{n+2},r_{n+1})
\end{aligned}$$

We are to show that $\mathcal{G}(r_{n+2},r_{n+2},r_{n+1})\le\mathcal{G}(r_{n+1},r_{n+1},r_n)$ for all $n\ge 0$. If not, then we have $n_0\ge 0$ with $\mathcal{G}(r_{n_0+1},r_{n_0+2},r_{n_0+2})>\mathcal{G}(r_{n_0},r_{n_0+1},r_{n_0+1})$. As $r_{n_0}\preceq r_{n_0+1}$, using inequality (6.8), we have

$$\begin{aligned}
&\psi\left(\mathcal{G}(r_{n_0+1},r_{n_0+2},r_{n_0+2})\right)\\
&= \psi\left(\mathcal{G}(fr_{n_0},fr_{n_0+1},fr_{n_0+1})\right)\\
&\le \psi\left(M(r_{n_0},r_{n_0+1},r_{n_0+1})\right)-\varphi\left(M(r_{n_0},r_{n_0+1},r_{n_0+1})\right)\\
&\le \psi\left((a_1+a_2+a_4)\mathcal{G}(r_{n_0},r_{n_0+1},r_{n_0+1})+(a_3+a_4)\mathcal{G}(r_{n_0+1},r_{n_0+2},r_{n_0+2})\right)\\
&\quad -\varphi\left(M(r_{n_0},r_{n_0+1},r_{n_0+1})\right)\\
&\le \psi((a_1+a_2+a_3+2a_4)\mathcal{G}(r_{n_0+1},r_{n_0+2},r_{n_0+2})-\varphi\left(M(r_{n_0},r_{n_0+1},r_{n_0+1})\right)
\end{aligned}$$

This implies that $\varphi\left(M(r_{n_0},r_{n_0+1},r_{n_0+1})\right)=0$, a contradiction. So for every $n\ge 0$,

$$\mathcal{G}(r_n,r_{n+1},r_{n+1})\ge\mathcal{G}(r_{n+1},r_{n+2},r_{n+2})$$

and as a result $M(r_{n+1},r_{n+1},r_n)\ \le\mathcal{G}(r_{n+1},r_{n+1},r_n)$. As $\{\mathcal{G}(r_{n+2},r_{n+2},r_{n+1})\}$ is a nonincreasing sequence, we have $\tau\ge 0$ such that $\lim_{n\to\infty}\mathcal{G}(r_{n+1},r_{n+2},r_{n+2})=\tau$. By lower semicontinuity of φ,

$$\varphi(a_1\tau)\le \liminf_{n\to\infty}\varphi\left(a_1\mathcal{G}(r_{n+1},r_{n+2},r_{n+2})\right)$$

Since $a_1\mathcal{G}(r_n,r_{n+1},r_{n+1})\le M(r_{n+1},r_{n+1},r_n)$,

$$\varphi(a_1\mathcal{G}(r_{n+2},r_{n+2},r_{n+1})\le\varphi\left(M(r_{n+1},r_{n+1},r_n)\right)$$

Now,

$$\begin{aligned}
\psi\left(\mathcal{G}(r_{n+1},r_{n+2},r_{n+2})\right) &\le \psi\left(M(r_n,r_{n+1},r_{n+1})\right)-\varphi\left(M(r_n,r_{n+1},r_{n+1})\right)\\
&\le \psi\left(\mathcal{G}(r_n,r_{n+1},r_{n+1})\right)-\varphi\left(M(r_n,r_{n+1},r_{n+1})\right)\\
&\le \psi\left(\mathcal{G}(r_n,r_{n+1},r_{n+1})\right)-\varphi\left(a_1\mathcal{G}(r_{n+1},r_{n+1},r_n)\right)
\end{aligned}$$

To show that $\tau=0$, we take the upper limits as $n\to\infty$ on the previous inequality to get

$$\begin{aligned}\psi(\tau) &\le \psi(\tau)-\liminf_{n\to\infty}\varphi\left(a_1\mathcal{G}\left(r_n,r_{n+1},r_{n+1}\right)\right)\\ &\le \psi(\tau)-\varphi(a_1\tau)\end{aligned}$$

This gives

$$\varphi(a_1\tau)\le 0$$

But then $\varphi(a_1\tau)=0$. Hence $\tau=0$, and

$$\lim_{n\to\infty}\mathcal{G}\left(r_{n+1},r_{n+2},r_{n+2}\right)=0 \tag{6.11}$$

Now, to claim $\{r_n\}$ is a $G-$Cauchy sequence in X. Assume on the contrary an $\delta>0$ exists, and n_k and m_k having $m_k>n_k>k$ implies

$$\mathcal{G}\left(r_{n_k},r_{m_k},r_{m_k}\right)\ge\delta \text{ and } \mathcal{G}\left(r_{n_k},r_{m_k-1},r_{m_k-1}\right)<\delta \tag{6.12}$$

Using (6.11) and (6.12),

$$\begin{aligned}\delta &\le \mathcal{G}\left(r_{n_k},r_{m_k},r_{m_k}\right)\\ &\le \mathcal{G}\left(r_{n_k},r_{m_k-1},r_{m_k-1}\right)+\mathcal{G}\left(r_{m_k-1},r_{m_k},r_{m_k}\right)\end{aligned}$$

implies

$$\lim_{k\to\infty}\mathcal{G}\left(r_{n_k},r_{m_k},r_{m_k}\right)=\delta \tag{6.13}$$

And

$$\begin{aligned}&\mathcal{G}\left(r_{n_k},r_{m_k},r_{m_k}\right)\\ &\le \mathcal{G}\left(r_{n_k},r_{m_k+1},r_{m_k+1}\right)+\mathcal{G}\left(r_{m_k+1},r_{m_k},r_{m_k}\right)\\ &\le \mathcal{G}\left(r_{n_k},r_{m_k+1},r_{m_k+1}\right)+\mathcal{G}\left(r_{m_k},r_{m_k+1},r_{m_k+1}\right)+\mathcal{G}\left(r_{m_k},r_{m_k+1},r_{m_k+1}\right)\end{aligned}$$

gives $\varepsilon\le\lim_{k\to\infty}\mathcal{G}\left(r_{n_k},r_{m_k+1},r_{m_k+1}\right)$.

Further,

$$\mathcal{G}\left(r_{n_k},r_{m_k+1},r_{m_k+1}\right)\le\mathcal{G}\left(r_{n_k},r_{m_k},r_{m_k}\right)+\mathcal{G}\left(r_{m_k},r_{m_k+1},r_{m_k+1}\right)$$

and using (6.11) and (6.13) with $\lim_{k\to\infty}\mathcal{G}\left(r_{n_k},r_{m_k+1},r_{m_k+1}\right)\le\delta$ implies

$$\lim_{k\to\infty}\mathcal{G}\left(r_{n_k},r_{m_k+1},r_{m_k+1}\right)=\delta \tag{6.14}$$

Now,

$$\begin{aligned}
&\mathcal{G}\left(r_{n_k}, r_{m_k}, r_{m_k}\right) \\
&\leq \mathcal{G}\left(r_{n_k}, r_{m_k+2}, r_{m_k+2}\right)+\mathcal{G}\left(r_{m_k+2}, r_{m_k}, r_{m_k}\right) \\
&\leq \mathcal{G}\left(r_{n_k}, r_{m_k+2}, r_{m_k+2}\right)+\mathcal{G}\left(r_{m_k}, r_{m_k+1}, r_{m_k+1}\right)+\mathcal{G}\left(r_{m_k}, r_{m_k+1}, r_{m_k+2}\right) \\
&\leq \mathcal{G}\left(r_{n_k}, r_{m_k+2}, r_{m_k+2}\right)+\mathcal{G}\left(r_{m_k}, r_{m_k+1}, r_{m_k+1}\right)+\mathcal{G}\left(r_{m_k}, r_{m_k+1}, r_{m_k+1}\right) \\
&\quad+\mathcal{G}\left(r_{m_k+1}, r_{m_k+1}, r_{m_k+2}\right) \\
&\leq \mathcal{G}\left(r_{n_k}, r_{m_k+2}, r_{m_k+2}\right)+\mathcal{G}\left(r_{m_k}, r_{m_k+1}, r_{m_k+1}\right)+\mathcal{G}\left(r_{m_k}, r_{m_k+1}, r_{m_k+1}\right) \\
&\quad+\mathcal{G}\left(r_{m_k+1}, r_{m_k+2}, r_{m_k+2}\right)+\mathcal{G}\left(r_{m_k+1}, r_{m_k+2}, r_{m_k+2}\right)
\end{aligned}$$

implies $\varepsilon \leq \lim_{k\to\infty} \mathcal{G}\left(r_{n_k}, r_{m_k+2}, r_{m_k+2}\right)$, and

$$\begin{aligned}
\mathcal{G}\left(r_{n_k}, r_{m_k+2}, r_{m_k+2}\right) &\leq \mathcal{G}\left(r_{n_k}, r_{m_k}, r_{m_k}\right)+\mathcal{G}\left(r_{m_k}, r_{m_k+2}, r_{m_k+2}\right) \\
&\leq \mathcal{G}\left(r_{n_k}, r_{m_k}, r_{m_k}\right)+\mathcal{G}\left(r_{m_k}, r_{m_k+1}, r_{m_k+1}\right) \\
&\quad+\mathcal{G}\left(r_{m_k+1}, r_{m_k+2}, r_{m_k+2}\right)
\end{aligned}$$

with (6.11) and (6.14) gives

$$\lim_{k\to\infty} \mathcal{G}\left(r_{n_k}, r_{m_k+2}, r_{m_k+2}\right) \leq \delta$$

So

$$\lim_{k\to\infty} \mathcal{G}\left(r_{n_k}, r_{m_k+2}, r_{m_k+2}\right) = \delta \tag{6.15}$$

By (6.13), with

$$\mathcal{G}\left(r_{n_k}, r_{m_k}, r_{m_k}\right) \leq \mathcal{G}\left(r_{n_k}, r_{m_k}, r_{m_k+1}\right)+\mathcal{G}\left(r_{m_k}, r_{m_k+1}, r_{m_k+1}\right)$$

we obtain $\varepsilon \leq \lim_{k\to\infty} \mathcal{G}\left(r_{n_k}, r_{m_k}, r_{m_k+1}\right)$.

But from

$$\begin{aligned}
\mathcal{G}\left(r_{n_k}, r_{m_k}, r_{m_k+1}\right) &\leq \mathcal{G}\left(r_{n_k}, r_{m_k}, r_{m_k}\right)+\mathcal{G}\left(r_{m_k}, r_{m_k}, r_{m_k+1}\right) \\
&\leq \mathcal{G}\left(r_{n_k}, r_{m_k}, r_{m_k}\right)+\mathcal{G}\left(r_{m_k}, r_{m_k+1}, r_{m_k+1}\right) \\
&\quad+\mathcal{G}\left(r_{m_k}, r_{m_k+1}, r_{m_k+1}\right)
\end{aligned}$$

with (6.11) and (6.13), it implies

$$\lim_{k\to\infty} \mathcal{G}\left(r_{n_k}, r_{m_k}, r_{m_k+1}\right) \leq \delta.$$

So

$$\lim_{k\to\infty} \mathcal{G}\left(r_{n_k}, r_{m_k}, r_{m_k+1}\right) = \delta \tag{6.16}$$

Now by definition of M and using all of (6.11), (6.14), (6.15), and (6.16), we can write

$$\begin{aligned}
&a_1\mathcal{G}\left(r_{n_k}, r_{m_k+1}, r_{m_k+1}\right) \leq M\left(r_{n_k}, r_{m_k+1}, r_{m_k+1}\right)\\
=\ & a_1\mathcal{G}\left(r_{n_k}, r_{m_k+1}, r_{m_k+1}\right) + a_2\mathcal{G}\left(r_{n_k}, fr_{n_k}, fr_{n_k}\right) + a_3\mathcal{G}\left(r_{m_k+1}, fr_{m_k+1}, fr_{m_k+1}\right)\\
&+a_4\left[\mathcal{G}\left(r_{n_k}, fr_{m_k+1}, fr_{m_k+1}\right) + \mathcal{G}\left(r_{m_k+1}, fr_{n_k}, fr_{n_k}\right)\right]\\
=\ & a_1\mathcal{G}\left(r_{n_k}, r_{m_k+1}, r_{m_k+1}\right) + a_2\mathcal{G}\left(r_{n_k}, r_{n_k+1}, r_{n_k+1}\right) + a_3\mathcal{G}\left(r_{m_k+1}, r_{m_k+2}, r_{m_k+2}\right)\\
&+a_4\left[\mathcal{G}\left(r_{n_k}, r_{m_k+2}, r_{m_k+2}\right) + \mathcal{G}\left(r_{m_k+1}, r_{n_k+1}, r_{n_k+1}\right)\right]\\
\leq\ & a_1\mathcal{G}\left(r_{n_k}, r_{m_k+1}, r_{m_k+1}\right) + a_2\mathcal{G}\left(r_{n_k}, r_{n_k+1}, r_{n_k+1}\right) + a_3\mathcal{G}\left(r_{m_k+1}, r_{m_k+2}, r_{m_k+2}\right)\\
&+a_4\left[\mathcal{G}\left(r_{n_k}, r_{m_k+2}, r_{m_k+2}\right) + \mathcal{G}\left(r_{n_k}, r_{n_k+1}, r_{n_k+1}\right) + \mathcal{G}\left(r_{n_k}, r_{m_k}, r_{m_k+1}\right)\right]
\end{aligned}$$

so that

$$\begin{aligned}
a_1\delta &\leq \lim_{k\to\infty} M\left(r_{n_k}, r_{m_k+1}, r_{m_k+1}\right) \leq a_1\delta + a_2(0) + a_3(0) + a_4[\delta+\delta]\\
&= (a_1 + 2a_4)\varepsilon \leq \delta
\end{aligned}$$

That is,

$$a_1\delta \leq \lim_{k\to\infty} M\left(r_{n_k}, r_{m_k+1}, r_{m_k+1}\right) \leq \delta$$

From (6.8),

$$\begin{aligned}
&\psi\left(\mathcal{G}\left(r_{n_k+1}, r_{m_k+2}, r_{m_k+2}\right)\right)\\
=\ &\psi\left(\mathcal{G}\left(fr_{n_k}, fr_{m_k+1}, fr_{m_k+1}\right)\right)\\
\leq\ &\psi\left(M\left(r_{n_k}, r_{m_k+1}, r_{m_k+1}\right)\right) - \varphi\left(M\left(r_{n_k}, r_{m_k+1}, r_{m_k+1}\right)\right)
\end{aligned}$$

and $k \to \infty$ gives

$$\psi(\delta) \leq \psi(\delta) - \varphi(a_1\delta)$$

which is not true, as $\delta > 0$.

Thus $\{r_n\}$ is a $G-$ Cauchy sequence. As X is $G-$ complete, an $\varpi \in X$ exists with $\{r_n\}$ $G-$ converging to ϖ as $n \to \infty$. In the case where f is $G-$ continuous, we simply obtain $f\varpi = \varpi$. When a $\{r_n\}$ nondecreasing sequence with $r_n \to \varpi$ in r implies $r_n \preceq \varpi$ for $n \geq 0$, by (6.8)

$$\begin{aligned}&\psi\left(\mathcal{G}\left(r_{n+1}, f\varpi, f\varpi\right)\right)\\&=\psi\left(\mathcal{G}\left(fr_n, f\varpi, f\varpi\right)\right)\leq\psi\left(M\left(r_n,\varpi,\varpi\right)\right)-\varphi\left(M\left(r_n,\varpi,\varpi\right)\right)\end{aligned}$$

with

$$\begin{aligned}&M\left(r_n,\varpi,\varpi\right)\\&=a_1\mathcal{G}\left(r_n,\varpi,\varpi\right)+a_2\mathcal{G}\left(fr_n, fr_n, r_n\right)+a_3\mathcal{G}\left(\varpi, f\varpi, f\varpi\right)\\&\quad+a_4\left[\mathcal{G}\left(r_n, f\varpi, f\varpi\right)+\mathcal{G}\left(\varpi, fr_n, fr_n\right)\right]\\&=a_1\mathcal{G}\left(r_n,\varpi,\varpi\right)+a_2\mathcal{G}\left(r_n, r_{n+1}, r_{n+1}\right)+a_3\mathcal{G}\left(\varpi, f\varpi, f\varpi\right)\\&\quad+a_4\left[\mathcal{G}\left(r_n, f\varpi, f\varpi\right)+\mathcal{G}\left(\varpi, r_{n+1}, r_{n+1}\right)\right]\end{aligned}$$

Thus $\lim_{n\to\infty} M\left(r_n,\varpi,\varpi\right)=\left(a_3+a_4\right)\mathcal{G}\left(f\varpi, f\varpi,\varpi\right)$.

Last,

$$\begin{aligned}&\psi\left(\mathcal{G}\left(\varpi, f\varpi, f\varpi\right)\right)\\&=\limsup_{n\to\infty}\psi\left(\mathcal{G}\left(fr_n, f\varpi, f\varpi\right)\right)\\&\leq\limsup_{n\to\infty}\left[\psi\left(M\left(r_n,\varpi,\varpi\right)\right)-\varphi\left(M\left(r_n,\varpi,\varpi\right)\right)\right]\\&\leq\psi\left(\left(a_3+a_4\right)\mathcal{G}\left(\varpi, f\varpi, f\varpi\right)\right)-\varphi\left(\left(a_3+a_4\right)\mathcal{G}\left(\varpi, f\varpi, f\varpi\right)\right)\\&\leq\psi\left(\mathcal{G}\left(\varpi, f\varpi, f\varpi\right)\right)-\varphi\left(\left(a_3+a_4\right)\mathcal{G}\left(\varpi, f\varpi, f\varpi\right)\right)\end{aligned}$$

implies $\varphi\left(\mathcal{G}\left(\varpi, f\varpi, f\varpi\right)\right)=0$, and, in turn, $\mathcal{G}\left(\varpi, f\varpi, f\varpi\right)=0$ implies $f\varpi=\varpi$, thereby completing the proof.

Example 6.5. Consider $X=[0,1]$ with $\mathcal{G}(r,s,r)=\max\{|r-s|,|s-r|,|r-r|\}$ given a G—metric on X. A map $f: X\to X$ defined as

$$f(r)=\begin{cases}\dfrac{r}{12} & \text{if } r\in[0,1)\\ \dfrac{1}{6} & \text{if } r=1\end{cases}$$

Consider $\varphi(s)=\frac{s}{10}$ *and* $\psi(s)=s$ *for* $s\in[0,\infty)$ *so that*

$$\psi\left(M(s,s,r)\right)-\varphi\left(M(s,s,r)\right)=\frac{9}{10}M(s,s,r)$$

For $r\preceq s$, *with* $r,s\in[0,1)$,

$$\mathcal{G}(r, fr, fr)=\frac{11}{12}r,\ \mathcal{G}(s, fs, fs)=\frac{11}{12}s$$

and

$$\frac{\mathcal{G}(r,fs,fs)+\mathcal{G}(s,fr,fr)}{2}=\frac{|12r-s|+12s-r}{24}$$

Now

$$\begin{aligned}
\mathcal{G}(fr,fs,fs) &= \frac{1}{12}(s-r)\\
&\le \frac{9}{10}\left\{\frac{1}{6}(s-r)+\frac{1}{6}\left(\frac{11r}{12}\right)+\frac{1}{6}\left(\frac{11s}{12}\right)+\frac{1}{6}\left(\frac{|12x-y|+12y-x}{24}\right)\right\}\\
&= \frac{9}{10}\{a_1\mathcal{G}(r,s,s)+a_2\mathcal{G}(r,fr,fr)+a_3\mathcal{G}(s,fs,fs)\\
&\quad +\frac{a_4}{2}\left[\mathcal{G}(r,fs,fs)+\mathcal{G}(s,fr,fr)\right]\}\\
&= \frac{9}{10}M(r,s,s)\\
&= \psi\left(M(r,s,s)\right)-\varphi\left(M(r,s,s)\right)
\end{aligned}$$

So (6.8) holds with $a_i=\frac{1}{6}$ *for* $i\in\{1,2,3,4\}$, *where* $a_1+a_2+a_3+2a_4\le 1.$

And where $r\in[0,1)$ *and* $s=1$,

$$\mathcal{G}(r,fr,fr)=\frac{11}{12}r,\ \mathcal{G}(s,fs,fs)=\frac{5}{6}$$

Now

$$\begin{aligned}
\mathcal{G}(fr,fs,fs) &= \frac{1}{12}(2-r)\\
&< \frac{9}{10}\left(\frac{1}{2}\right)\\
&= \frac{9}{10}a_3\mathcal{G}(s,fs,fs)\\
&\le \frac{9}{10}M(r,s,s)\\
&= \psi\left(M(r,s,s)\right)-\varphi\left(M(r,s,s)\right)
\end{aligned}$$

So (6.8) holds with $a_1=a_2=a_4=0$ *and* $a_3=\frac{3}{5}$ *where* $a_1+a_2+a_3+2a_4\le 1$.

Hence all axioms of Theorem 6.2 hold and $0\in F(f)$.

Corollary 6.3. *In a complete* $G-$ *metric space* $(X,\preceq,\mathcal{G})$, *if a self map* f *on* X *is nondecreasing and for all comparable* $r,s\in X$,

$$\mathcal{G}(fr,fs,fs)\le M(r,s,s)-\varphi\left(M(r,s,s)\right) \tag{6.17}$$

holds for $\varphi \in \Phi$ *and*

$$M(r,s,s) = a_1\mathcal{G}(r,s,s) + a_2\mathcal{G}(r,fr,fr) + a_3\mathcal{G}(s,fs,fs) + a_4\left[\mathcal{G}(r,fs,fs) + \mathcal{G}(s,fr,fr)\right]$$

for $0 < a_i$ *for all* $i = 1,\dots,4$ *with* $\sum_{i=1}^{4} a_i < 1$. *If there is an* $r_0 \in X$ *with* $fr_0 \succeq r_0$ *and either* f *is G-continuous or a nondecreasing sequence* $\{r_n\}$ *with* $r_n \to z$ *in* X *implies* $z_n \preceq z$ *for all* $n \in \mathbb{N}$, *then* f *has at most one fixed point.*

Theorem 6.3. *In a complete* $G-$ *metric space* $(X,\preceq,\mathcal{G})$, *if a self map* f *on* X *is nondecreasing and for all* $r,s \in X$ *having* $y \succeq x$,

$$\psi\left(\mathcal{G}(fr,fs,fu)\right) \le \psi\left(M(r,s,u)\right) - \varphi\left(M(r,s,u)\right) \tag{6.18}$$

holds for $\varphi \in \Phi$ *and*

$$M(r,s,s) = \max\left\{\mathcal{G}(r,s,s),\mathcal{G}(fr,fr,r),\mathcal{G}(fs,fs,s),\left[\mathcal{G}(r,fs,fs) + \mathcal{G}(s,fr,fr)\right]\right\}$$

If there is an $r_0 \in X$ *with* $fr_0 \succeq r_0$ *and either* f *is G-continuous or a nondecreasing sequence* $\{r_n\}$ *with* $r_n \to z$ *in* X *gives* $z_n \preceq z$ *for all* $n \in \mathbb{N}$, *then* f *has at most one fixed point.*

Proof. If $fr_0 = r_0$, we are done. Suppose that $fr_0 \neq r_0$. Now by using nondecreasingness of f with $fr_0 \succeq r_0$, we get

$$r_0 \preceq fr_0 \preceq f^2 r_0 \preceq \dots \preceq f^n r_0 \preceq \dots$$

Define $\{r_n\}$ with $r_n = f^n r_0$ and $r_{n+1} = fr_n$. We are considering $G(r_n, r_{n+1}, r_{n+1}) > 0$, for every $n \ge 0$. In that case, $r_l = r_{l+1}$ for any l. For those l, using (6.18), we have

$$\begin{aligned} &\psi\left(\mathcal{G}(r_{n+1}, r_{n+2}, r_{n+2})\right) \\ &= \psi\left(\mathcal{G}(fr_n, fr_{n+1}, fr_{n+1})\right) \\ &\le \psi\left(M(r_n, r_{n+1}, r_{n+1})\right) - \varphi\left(M(r_n, r_{n+1}, r_{n+1})\right) \end{aligned} \tag{6.19}$$

where

$$\begin{aligned} &M(r_n, r_{n+1}, r_{n+1}) \\ &= \max\left\{\mathcal{G}(r_n, r_{n+1}, r_{n+1}), \mathcal{G}(r_n, fr_n, fr_n), \mathcal{G}(r_{n+1}, fr_{n+1}, fr_{n+1}), \left[\mathcal{G}(r_n, fr_{n+1}, fr_{n+1}) + \mathcal{G}(r_{n+1}, fr_n, fr_n)\right]/2\right\} \\ &= \max\left\{\mathcal{G}(r_n, r_{n+1}, r_{n+1}), \mathcal{G}(r_n, r_{n+1}, r_{n+1}), \mathcal{G}(r_{n+1}, r_{n+2}, r_{n+2}),\right. \end{aligned}$$

$$\left[\mathcal{G}\left(r_n, r_{n+2}, r_{n+2}\right)+\mathcal{G}\left(r_{n+1}, r_{n+1}, r_{n+1}\right)\right]/2\Big\}$$
$$= \mathcal{G}\left(r_{n+1}, r_{n+2}, r_{n+2}\right)$$

Hence

$$\psi\left(\mathcal{G}\left(r_{n+1}, r_{n+2}, r_{n+2}\right)\right) \leq \psi\left(\mathcal{G}\left(r_{n+1}, r_{n+2}, r_{n+2}\right)\right)-\varphi\left(\mathcal{G}\left(r_{n+1}, r_{n+2}, r_{n+2}\right)\right)$$

implies that $\varphi\left(\mathcal{G}\left(r_{n+1}, r_{n+2}, r_{n+2}\right)\right)=0$, and so $r_{n+1}=r_{n+2}$. Thus $r_{n+1}=f\left(r_{n+1}\right)$.

Now, by taking $\mathcal{G}\left(r_{n+1}, r_{n+1}, r_n\right)>0$ for $n \geq 0$,

$$\begin{aligned} & \psi\left(\mathcal{G}\left(r_{n+1}, r_{n+2}, r_{n+2}\right)\right) \\ = & \psi\left(\mathcal{G}\left(f r_n, f r_{n+1}, f r_{n+1}\right)\right) \\ \leq & \psi\left(M\left(r_n, r_{n+1}, r_{n+1}\right)\right)-\varphi\left(M\left(r_n, r_{n+1}, r_{n+1}\right)\right) \end{aligned} \tag{6.20}$$

where

$$\begin{aligned} & M\left(r_n, r_{n+1}, r_{n+1}\right) \\ = & \max \Big\{\mathcal{G}\left(r_n, r_{n+1}, r_{n+1}\right), \mathcal{G}\left(r_n, f r_n, f r_n\right), \mathcal{G}\left(r_{n+1}, f r_{n+1}, f r_{n+1}\right) \\ & \left[\mathcal{G}\left(r_n, f r_{n+1}, f r_{n+1}\right)+\mathcal{G}\left(r_{n+1}, f r_n, f r_n\right)\right]/2\Big\} \\ = & \max \Big\{\mathcal{G}\left(r_n, r_{n+1}, r_{n+1}\right), \mathcal{G}\left(r_n, r_{n+1}, r_{n+1}\right), \mathcal{G}\left(r_{n+1}, r_{n+2}, r_{n+2}\right) \\ & \left[\mathcal{G}\left(r_n, r_{n+2}, r_{n+2}\right)+\mathcal{G}\left(r_{n+1}, r_{n+1}, r_{n+1}\right)\right]/2\Big\} \\ \leq & \max \Big\{\mathcal{G}\left(r_n, r_{n+1}, r_{n+1}\right), \mathcal{G}\left(r_{n+1}, r_{n+2}, r_{n+2}\right) \\ & \left[\mathcal{G}\left(r_n, r_{n+1}, r_{n+1}\right)+\mathcal{G}\left(r_{n+1}, r_{n+2}, r_{n+2}\right)\right]/2\Big\} \\ = & \max \left\{\mathcal{G}\left(r_n, r_{n+1}, r_{n+1}\right), \mathcal{G}\left(r_{n+1}, r_{n+2}, r_{n+2}\right)\right\} \end{aligned}$$

In the case where $\mathcal{G}\left(r_{n+1}, r_{n+2}, r_{n+2}\right)>\mathcal{G}\left(r_n, r_{n+1}, r_{n+1}\right)$ for any $n \geq 0$, it gives $\varphi\left(M\left(r_n, r_{n+1}, r_{n+1}\right)\right)=0$, a contradiction. Thus for $n \geq 0$,

$$\mathcal{G}\left(r_{n+2}, r_{n+2}, r_{n+1}\right) \leq \mathcal{G}\left(r_{n+1}, r_n, r_{n+1}\right)$$

So $\left\{\mathcal{G}\left(r_{n+2}, r_{n+2}, r_{n+1}\right)\right\}$ is a nonincreasing sequence, and we have $\tau \geq 0$ with $\lim_{\imath \to \infty} \mathcal{G}\left(r_{n+2}, r_{n+1}, r_{n+2}\right)=\tau$. This gives $\lim_{n \to \infty} \mathcal{G}\left(r_n, r_{n+1}, r_{n+1}\right)=\lim_{n \to \infty} M\left(r_n, r_{n+1}, r_{n+1}\right)=\tau$. As

$$\varphi(\tau) \leq \liminf_{n \to \infty} \varphi\left(M\left(r_n, r_{n+1}, r_{n+1}\right)\right)$$

to show that $\tau = 0$. By upper limit as $n \to \infty$ to

$$\psi\left(\mathcal{G}\left(r_{n+2}, r_{n+2}, r_{n+1}\right)\right) \leq \psi\left(M\left(r_{n+1}, r_{n+1}, r_{n}\right)\right) - \varphi\left(M\left(r_{n+1}, r_{n+1}, r_{n}\right)\right)$$

and so

$$\begin{aligned} \psi(\tau) &\leq \psi(\tau) - \liminf_{n\to\infty} \varphi\left(M\left(r_{n}, r_{n+1}, r_{n+1}\right)\right) \\ &\leq \psi(\tau) - \varphi(\tau) \end{aligned}$$

This implies $\varphi(\tau) = 0$, and so

$$\lim_{n\to\infty} \mathcal{G}\left(r_{n+1}, r_{n+2}, r_{n+2}\right) = 0 \quad (6.21)$$

Now, to prove $\{r_n\}$ is a $G-$ Cauchy sequence in X. If not, there is $\varepsilon > 0$, and there exist integers n_k and m_k with $m_k > n_k > k$ such that

$$\mathcal{G}\left(r_{n_k}, r_{m_k}, r_{m_k}\right) \geq \varepsilon \text{ and } \mathcal{G}\left(r_{n_k}, r_{m_k-1}, r_{m_k-1}\right) < \varepsilon \quad (6.22)$$

Next,

$$\begin{aligned} \varepsilon &\leq \mathcal{G}\left(r_{n_k}, r_{m_k}, r_{m_k}\right) \\ &\leq \mathcal{G}\left(r_{n_k}, r_{m_k-1}, r_{m_k-1}\right) + \mathcal{G}\left(r_{m_k-1}, r_{m_k}, r_{m_k}\right) \end{aligned}$$

gives by (6.21) and (6.22) that

$$\lim_{k\to\infty} \mathcal{G}\left(r_{n_k}, r_{m_k}, r_{m_k}\right) = \varepsilon \quad (6.23)$$

And

$$\begin{aligned} &\mathcal{G}\left(r_{n_k}, r_{m_k}, r_{m_k}\right) \\ &\leq \mathcal{G}\left(r_{n_k}, r_{m_k+1}, r_{m_k+1}\right) + \mathcal{G}\left(r_{m_k+1}, r_{m_k}, r_{m_k}\right) \\ &\leq \mathcal{G}\left(r_{n_k}, r_{m_k+1}, r_{m_k+1}\right) + \mathcal{G}\left(r_{m_k}, r_{m_k+1}, r_{m_k+1}\right) + \mathcal{G}\left(r_{m_k}, r_{m_k+1}, r_{m_k+1}\right) \end{aligned}$$

gives $\varepsilon \leq \lim_{k\to\infty} \mathcal{G}\left(r_{n_k}, r_{m_k+1}, r_{m_k+1}\right)$. An application of (6.21) together with (6.23) implies

$$\mathcal{G}\left(r_{n_k}, r_{m_k+1}, r_{m_k+1}\right) \leq \mathcal{G}\left(r_{n_k}, r_{m_k}, r_{m_k}\right) + \mathcal{G}\left(r_{m_k}, r_{m_k+1}, r_{m_k+1}\right)$$

so that $\lim_{k\to\infty} \mathcal{G}\left(r_{n_k}, r_{m_k+1}, r_{m_k+1}\right) \leq \varepsilon$ and hence

$$\lim_{k\to\infty} \mathcal{G}\left(r_{n_k}, r_{m_k+1}, r_{m_k+1}\right) = \varepsilon \quad (6.24)$$

Stepping forward,

$$\begin{aligned}
&\mathcal{G}\left(r_{n_k}, r_{m_k}, r_{m_k}\right) \\
&\leq \mathcal{G}\left(r_{n_k}, r_{m_k+2}, r_{m_k+2}\right)+\mathcal{G}\left(r_{m_k+2}, r_{m_k}, r_{m_k}\right) \\
&\leq \mathcal{G}\left(r_{n_k}, r_{m_k+2}, r_{m_k+2}\right)+\mathcal{G}\left(r_{m_k}, r_{m_k+1}, r_{m_k+1}\right)+\mathcal{G}\left(r_{m_k}, r_{m_k+1}, r_{m_k+2}\right) \\
&\leq \mathcal{G}\left(r_{n_k}, r_{m_k+2}, r_{m_k+2}\right)+\mathcal{G}\left(r_{m_k}, r_{m_k+1}, r_{m_k+1}\right)+\mathcal{G}\left(r_{m_k}, r_{m_k+1}, r_{m_k+1}\right) \\
&+\mathcal{G}\left(r_{m_k+1}, r_{m_k+1}, r_{m_k+2}\right) \\
&\leq \mathcal{G}\left(r_{n_k}, r_{m_k+2}, r_{m_k+2}\right)+\mathcal{G}\left(r_{m_k}, r_{m_k+1}, r_{m_k+1}\right)+\mathcal{G}\left(r_{m_k}, r_{m_k+1}, r_{m_k+1}\right) \\
&+\mathcal{G}\left(r_{m_k+1}, r_{m_k+2}, r_{m_k+2}\right)+\mathcal{G}\left(r_{m_k+1}, r_{m_k+2}, r_{m_k+2}\right)
\end{aligned}$$

implies $\varepsilon \leq \lim_{k \to \infty} \mathcal{G}\left(r_{n_k}, r_{m_k+2}, r_{m_k+2}\right)$, and

$$\begin{aligned}
\mathcal{G}\left(r_{n_k}, r_{m_k+2}, r_{m_k+2}\right) &\leq \mathcal{G}\left(r_{n_k}, r_{m_k}, r_{m_k}\right)+\mathcal{G}\left(r_{m_k}, r_{m_k+2}, r_{m_k+2}\right) \\
&\leq \mathcal{G}\left(r_{n_k}, r_{m_k}, r_{m_k}\right)+\mathcal{G}\left(r_{m_k}, r_{m_k+1}, r_{m_k+1}\right) \\
&+\mathcal{G}\left(r_{m_k+1}, r_{m_k+2}, r_{m_k+2}\right)
\end{aligned}$$

Also, exploiting (6.21) and (6.24) yields

$$\lim_{k \to \infty} \mathcal{G}\left(r_{n_k}, r_{m_k+2}, r_{m_k+2}\right) \leq \varepsilon$$

As a result,

$$\lim_{k \to \infty} \mathcal{G}\left(r_{n_k}, r_{m_k+2}, r_{m_k+2}\right) = \varepsilon \tag{6.25}$$

Furthermore,

$$\mathcal{G}\left(r_{n_k}, r_{m_k}, r_{m_k}\right) \leq \mathcal{G}\left(r_{n_k}, r_{m_k}, r_{m_k+1}\right)+\mathcal{G}\left(r_{m_k}, r_{m_k+1}, r_{m_k+1}\right)$$

yields $\varepsilon \leq \lim_{k \to \infty} \mathcal{G}\left(r_{n_k}, r_{m_k}, r_{m_k+1}\right)$ by (6.21) and (6.23).

On the other hand,

$$\begin{aligned}
\mathcal{G}\left(r_{n_k}, r_{m_k}, r_{m_k+1}\right) &\leq \mathcal{G}\left(r_{n_k}, r_{m_k}, r_{m_k}\right)+\mathcal{G}\left(r_{m_k}, r_{m_k}, r_{m_k+1}\right) \\
&\leq \mathcal{G}\left(r_{n_k}, r_{m_k}, r_{m_k}\right)+\mathcal{G}\left(r_{m_k}, r_{m_k+1}, r_{m_k+1}\right) \\
&+\mathcal{G}\left(r_{m_k}, r_{m_k+1}, r_{m_k+1}\right)
\end{aligned}$$

Another application of (6.21) and (6.23),

$$\lim_{k\to\infty} \mathcal{G}\left(r_{n_k}, r_{m_k}, r_{m_k+1}\right) \leq \varepsilon$$

and hence,

$$\lim_{k\to\infty} \mathcal{G}\left(r_{n_k}, r_{m_k}, r_{m_k+1}\right) = \varepsilon \tag{6.26}$$

From the definition of M and all of (6.21), (6.24), (6.25), and (6.26), we have

$$\begin{aligned}
&\mathcal{G}\left(r_{n_k}, r_{m_k+1}, r_{m_k+1}\right)\\
&\leq M\left(r_{n_k}, r_{m_k+1}, r_{m_k+1}\right)\\
&= \max\Big\{\mathcal{G}\left(r_{n_k}, r_{m_k+1}, r_{m_k+1}\right), \mathcal{G}\left(r_{n_k}, fr_{n_k}, fr_{n_k}\right), \mathcal{G}\left(r_{m_k+1}, fr_{m_k+1}, fr_{m_k+1}\right)\\
&\quad \left[\mathcal{G}\left(r_{n_k}, fr_{m_k+1}, fr_{m_k+1}\right) + \mathcal{G}\left(r_{m_k+1}, fr_{n_k}, fr_{n_k}\right)\right]/2\Big\}\\
&= \max\Big\{\mathcal{G}\left(r_{n_k}, r_{m_k+1}, r_{m_k+1}\right), \mathcal{G}\left(r_{n_k}, r_{n_k+1}, r_{n_k+1}\right), \mathcal{G}\left(r_{m_k+1}, r_{m_k+2}, r_{m_k+2}\right)\\
&\quad \left[\mathcal{G}\left(r_{n_k}, r_{m_k+2}, r_{m_k+2}\right) + \mathcal{G}\left(r_{m_k+1}, r_{n_k+1}, r_{n_k+1}\right)\right]/2\Big\}\\
&\leq \max\Big\{\mathcal{G}\left(r_{n_k}, r_{m_k+1}, r_{m_k+1}\right), \mathcal{G}\left(r_{n_k}, r_{n_k+1}, r_{n_k+1}\right), \mathcal{G}\left(r_{m_k+1}, r_{m_k+2}, r_{m_k+2}\right)\\
&\quad \left[\mathcal{G}\left(r_{n_k}, r_{m_k+2}, r_{m_k+2}\right) + \mathcal{G}\left(r_{n_k}, r_{n_k+1}, r_{n_k+1}\right) + \mathcal{G}\left(r_{n_k}, r_{m_k}, r_{m_k+1}\right)\right]/2\Big\}
\end{aligned}$$

This gives

$$\delta \leq \lim_{k\to\infty} M\left(r_{n_k}, r_{m_k+1}, r_{m_k+1}\right) \leq \max\left\{\delta, 0, \delta, \left[\delta+\delta\right]/2\right\} = \delta$$

and so

$$\lim_{k\to\infty} M\left(r_{n_k}, r_{m_k+1}, r_{m_k+1}\right) = \delta$$

From (6.21),

$$\begin{aligned}
&\psi\left(\mathcal{G}\left(r_{n_k+1}, r_{m_k+2}, r_{m_k+2}\right)\right)\\
&= \psi\left(\mathcal{G}\left(fr_{n_k}, fr_{m_k+1}, fr_{m_k+1}\right)\right)\\
&\leq \psi\left(M\left(r_{n_k}, r_{m_k+1}, r_{m_k+1}\right)\right) - \varphi\left(M\left(r_{n_k}, r_{m_k+1}, r_{m_k+1}\right)\right)
\end{aligned}$$

and by the limit $k \to \infty$,

$$\psi(\delta) \leq \psi(\delta) - \varphi(\delta)$$

a contradiction, as $\delta > 0$.

We obtain that $\{r_n\}$ is a $G-$ Cauchy sequence, so $\varpi \in X$ exists for which $\{r_n\}$ $G-$ converges to ϖ. By continuity of f, we simply get $f\varpi = \varpi$. The case of a nondecreasing sequence $\{r_n\}$ with $r_n \to \varpi$ implies $r_n \preceq \varpi$ for $n \geq 0$, and from (6.18)

$$\begin{aligned} &\psi\left(\mathcal{G}\left(r_{n+1}, f\varpi, f\varpi\right)\right) \\ = &\ \psi\left(\mathcal{G}\left(fr_n, f\varpi, f\varpi\right)\right) \leq \psi\left(M\left(r_n, \varpi, \varpi\right)\right) - \varphi\left(M\left(r_n, \varpi, \varpi\right)\right) \end{aligned} \tag{6.27}$$

where

$$\begin{aligned} &M\left(r_n, \varpi, \varpi\right) \\ = &\ \max\left\{\mathcal{G}\left(r_n, \varpi, \varpi\right), \mathcal{G}\left(r_n, fr_n, fr_n\right), \mathcal{G}\left(\varpi, f\varpi, f\varpi\right)\right. \\ &\left.\left[\mathcal{G}\left(r_n, f\varpi, f\varpi\right) + \mathcal{G}\left(\varpi, fr_n, fr_n\right)\right]/2\right\} \\ = &\ \max\left\{\mathcal{G}\left(r_n, \varpi, \varpi\right), \mathcal{G}\left(r_n, r_{n+1}, r_{n+1}\right), \mathcal{G}\left(\varpi, f\varpi, f\varpi\right)\right. \\ &\left.\left[\mathcal{G}\left(r_n, f\varpi, f\varpi\right) + \mathcal{G}\left(\varpi, r_{n+1}, r_{n+1}\right)\right]/2\right\} \end{aligned}$$

So $\lim_{n\to\infty} M\left(r_n, \varpi, \varpi\right) = \mathcal{G}\left(\varpi, f\varpi, f\varpi\right)$. Thus from (6.27), we obtain

$$\begin{aligned} &\psi\left(\mathcal{G}\left(\varpi, f\varpi, f\varpi\right)\right) \\ = &\ \limsup_{n\to\infty} \psi\left(\mathcal{G}\left(fr_n, f\varpi, f\varpi\right)\right) \leq \limsup_{n\to\infty}\left[\psi\left(M\left(r_n, \varpi, \varpi\right)\right) - \varphi\left(M\left(r_n, \varpi, \varpi\right)\right)\right] \\ \leq &\ \psi\left(\mathcal{G}\left(f\varpi, f\varpi, \varpi\right)\right) - \varphi\left(\mathcal{G}\left(f\varpi, f\varpi, \varpi\right)\right) \end{aligned}$$

This gives $\varphi\left(\mathcal{G}\left(f\varpi, f\varpi, \varpi\right)\right) = 0$ so that $\mathcal{G}\left(\varpi, f\varpi, f\varpi\right) = 0$ and hence $f\varpi = \varpi$.

Corollary 6.4. *In a complete $G-$ metric space $\left(X, \preceq, \mathcal{G}\right)$, if a self map f on X is nondecreasing and for all $r, s \in X$ with $r \preceq s$,*

$$G(fr, fs, fs) \leq M(r, s, s) - \varphi\left(M(r, s, s)\right) \tag{6.28}$$

holds for $\varphi \in \Phi$, and

$$M\left(r, s, s\right) = \max\left\{\mathcal{G}\left(r, s, s\right), \mathcal{G}\left(r, fr, fr\right), \mathcal{G}\left(s, fs, fs\right)\right. \left.\frac{\mathcal{G}\left(r, fs, fs\right) + \mathcal{G}\left(s, fr, fr\right)]}{2}\right\}$$

If there is an $r_0 \in X$ with $fr_0 \succeq r_0$ and either f is G-continuous or a nondecreasing sequence $\{r_n\}$ with $r_n \to z$ in X implies $z_n \preceq z$ for all $n \in \mathbb{N}$, then f has at most one fixed point.

Corollary 6.5. *In a complete $G-$ metric space $(X,\preceq,\mathcal{G})$, if a self map f on X is nondecreasing and for all $r,s\in X$ with $r\preceq s$,*

$$\mathcal{G}(fs,fs,fr) \leq k\max\{\mathcal{G}(r,s,s),\mathcal{G}(s,fs,fs),\mathcal{G}(r,fr,fr) \quad [\mathcal{G}(r,fs,fs)+\mathcal{G}(s,fr,fr)]/2\} \tag{6.29}$$

where $k\in[0,1)$. If there is an $r_0\in X$ with $fr_0\succeq r_0$ and either f is G-continuous or a nondecreasing sequence $\{r_n\}$ with $r_n\to z$ in X implies $z_n\preceq z$ for all $n\in\mathbb{N}$, then f has at most one fixed point.

Proof. Define $\varphi,\psi:[0,\infty)\to[0,\infty)$ by $\psi(\lambda)=\lambda$ and $\varphi(\lambda)=(1-k)\lambda$ for $\lambda\in[0,\infty)$, where $k\in[0,1)$. Then we get $\psi\in\Psi$ and $\varphi\in\Phi$. Follow the proof by using Theorem 6.3.

Corollary 6.6. *In a complete $G-$ metric space $(X,\preceq,\mathcal{G})$, if a self map f on X is nondecreasing and for all $r,s\in X$ with $r,s\in X$*

$$\psi(\mathcal{G}(fr,fs,fs))\leq\psi(\mathcal{G}(r,s,s))-\varphi(\mathcal{G}(r,s,s)) \tag{6.30}$$

holds for $\varphi\in\Phi$, $\psi\in\Psi$. If there is an $r_0\in X$ with $fr_0\succeq r_0$ and either f is G-continuous or a nondecreasing sequence $\{r_n\}$ with $r_n\to z$ in X gives $z_n\preceq z$ for all $n\in\mathbb{N}$, then f has at most one fixed point.

Corollary 6.7. *In a complete ordered $G-$ metric space $(X,\preceq,\mathcal{G})$, if a self map f on X is nondecreasing and for all $s,r\in X$ with $y\succeq x$,*

$$\mathcal{G}(fr,fs,fs)\leq\frac{\mathcal{G}(r,s,s)}{1+\mathcal{G}(r,s,s)} \tag{6.31}$$

holds for $\psi\in\Psi$. If there is an $r_0\in X$ with $fr_0\succeq r_0$ and either f is G-continuous or a nondecreasing sequence $\{r_n\}$ with $r_n\to z$ in X gives $r_n\preceq z$ for all $n\in\mathbb{N}$, then f has at most one fixed point.

Proof. Define $\varphi,\psi:[0,\infty)\to[0,\infty)$ by $\psi(\lambda)=\lambda$ and $\varphi(\lambda)=\dfrac{\lambda^2}{1+\lambda}$ for $\lambda\in[0,\infty)$. Then we get $\psi\in\Psi$ and $\varphi\in\Phi$. Follow the proof by using Corollary 6.6.

6.5 *T*–HARDY ROGERS SELF-MAPS IN ORDERED *G*–METRIC SPACES

Abbas and Nazir [13] studied $T-$ Hardy Rogers–type contractions for the fixed points of self maps. We are establishing some new fixed point results of general $T-$ Hardy Rogers–type contractive maps in the ordered $G-$ metric space structure.

Theorem 6.4. *In a complete ordered $G-$ metric space $(X,\preceq,\mathcal{G})$, if $T:X\to X$ is an injective and continuous map and $f:X\to X$ is nondecreasing and for all comparable $r,s,u\in X$,*

$$\begin{aligned}
&\mathcal{G}(Tfr,Tfs,Tfu) \\
&\le \alpha\mathcal{G}(Tr,Ts,Tu)+\beta\left[\mathcal{G}(Tr,Tfr,Tfr)+\mathcal{G}(Ts,Tfs,Tfs)+\mathcal{G}(Tu,Tfu,Tfu)\right] \\
&\quad +\gamma\left[\mathcal{G}(Tfr,Tfr,Ts)+\mathcal{G}(Tfs,Tfs,Tu)+\mathcal{G}(Tfu,Tfu,Tr)\right]
\end{aligned} \tag{6.32}$$

holds for $\alpha,\beta,\gamma \ge 0$ with $\alpha+3\beta+3\gamma<1$. If there is an $r_0 \in X$ with $fr_0 \succeq r_0$ and either f is G-continuous or a nondecreasing sequence $\{r_n\}$ with $r_n \to z$ in X gives $z_n \preceq z$ for all $n \in \mathbb{N}$, then set $F(f)$ is non-empty provided that T is subsequentially convergent. Moreover, $F(f)$ has at most one element if and only if $F(f)$ is well ordered.

Proof. By nondecreasing of f, it implies

$$r_1 = fr_0 \preceq f^2 r_0 \preceq \ldots \preceq f^n r_0 \preceq f^{n+1} r_0 \preceq \ldots$$

Define a sequence $\{r_n\}$ in X with $r_n = f^n r_0$ i.e., $r_{n+1} = fr_n$ for $n \ge 0$. If for any k, $r_k = fr_k$, then the result follows. Let us take $r_n \ne fr_n$ for each n, as $r_n \preceq r_{n+1}$ for every n. By (6.32),

$$\begin{aligned}
&\mathcal{G}(Tr_{n+1},Tr_{n+2},Tr_{n+3}) \\
&= \mathcal{G}(Tfr_n,Tfr_{n+1},Tfr_{n+2}) \\
&\le \alpha\mathcal{G}(Tr_n,Tr_{n+1},Tr_{n+2})+\beta[\mathcal{G}(Tr_n,Tfr_n,Tfr_n)+\mathcal{G}(Tr_{n+1},Tfr_{n+1},Tfr_{n+1}) \\
&\quad +\mathcal{G}(Tr_{n+2},Tfr_{n+2},Tfr_{n+2})]+\gamma[\mathcal{G}(Tfr_n,Tfr_n,Tr_{n+1})+\mathcal{G}(Tfr_{n+1},Tfr_{n+1},Tr_n) \\
&\quad +\mathcal{G}(Tfr_{n+2},Tfr_{n+2},Tr_n)] \\
&= \alpha\mathcal{G}(Tr_n,Tr_{n+1},Tr_{n+2})+\beta[\mathcal{G}(Tr_n,Tr_{n+1},Tr_{n+1})+\mathcal{G}(Tr_{n+1},Tr_{n+2},Tr_{n+2}) \\
&\quad +\mathcal{G}(Tr_{n+2},Tr_{n+3},Tr_{n+3})]+\gamma[\mathcal{G}(Tr_{n+1},Tr_{n+1},Tr_{n+1})+\mathcal{G}(Tr_{n+2},Tr_{n+2},Tr_n) \\
&\quad +\mathcal{G}(Tr_{n+3},Tr_{n+3},Tr_n)] \\
&\le \alpha\mathcal{G}(Tr_n,Tr_{n+1},Tr_{n+2})+\beta[\mathcal{G}(Tr_n,Tr_{n+1},Tr_{n+2})+\mathcal{G}(Tr_{n+1},Tr_{n+2},Tr_{n+3}) \\
&\quad +\mathcal{G}(Tr_{n+2},Tr_{n+3},Tr_{n+1})]+\gamma[\mathcal{G}(Tr_{n+2},Tr_{n+1},Tr_n) \\
&\quad +\mathcal{G}(Tr_{n+3},Tr_{n+3},Tr_{n+2})+\mathcal{G}(Tr_{n+2},Tr_{n+2},Tr_n)] \\
&\le (\alpha+\beta)\mathcal{G}(Tr_n,Tr_{n+1},Tr_{n+2})+2\beta\mathcal{G}(Tr_{n+1},Tr_{n+2},Tr_{n+3}) \\
&\quad +\gamma\left[\mathcal{G}(Tr_n,Tr_{n+1},Tr_{n+2})+\mathcal{G}(Tr_{n+3},Tr_{n+2},Tr_{n+1})+\mathcal{G}(Tr_{n+2},Tr_{n+1},Tr_n)\right] \\
&= (\alpha+\beta+2\gamma)\mathcal{G}(Tr_n,Tr_{n+1},Tr_{n+2})+(2\beta+\gamma)\mathcal{G}(Tr_{n+1},Tr_{n+2},Tr_{n+3})
\end{aligned}$$

that is,

$$(1-2\beta-\gamma)\mathcal{G}(Tr_{n+1},Tr_{n+2},Tr_{n+3}) \leq (\alpha+\beta+2\gamma)\mathcal{G}(Tr_n,Tr_{n+1},Tr_{n+2})$$

Hence,

$$\mathcal{G}(Tr_{n+3},Tr_{n+2},Tr_{n+1}) \leq \lambda\mathcal{G}(Tr_{n+2},Tr_{n+1},Tr_{n+1}) \tag{6.33}$$

with $\lambda = \dfrac{\alpha+\beta+2\gamma}{1-(2\beta+\gamma)} < 1$. Now

$$\begin{aligned}\mathcal{G}(Tr_{n+1},Tr_{n+2},Tr_{n+3}) &\leq \lambda\mathcal{G}(Tr_n,Tr_{n+1},Tr_{n+2})\\ &\leq \cdots \leq \lambda^{n+1}\mathcal{G}(Tr_0,Tr_1,Tr_2)\end{aligned}$$

For $n < m$,

$$\begin{aligned}\mathcal{G}(Tr_n,Tr_m,Tr_m) &\leq \mathcal{G}(Tr_n,Tr_{n+1},Tr_{n+1}) + \mathcal{G}(Tr_{n+1},Tr_{n+2},Tr_{n+2})\\ &\quad + \cdots + \mathcal{G}(Tr_{m-1},Tr_m,Tr_m)\\ &\leq \mathcal{G}(Tr_n,Tr_{n+1},Tr_{n+2}) + \mathcal{G}(Tr_{n+1},Tr_{n+2},Tr_{n+3})\\ &\quad + \cdots + \mathcal{G}(Tr_{m-2},Tr_{m-1},Tr_m)\\ &\leq \left[\lambda^n + \lambda^{n+1} + \cdots + \lambda^{m-1}\right]\mathcal{G}(Tr_0,Tr_1,Tr_2)\\ &\leq \frac{\lambda^n}{1-\lambda}\mathcal{G}(Tr_0,Tr_1,Tr_2)\end{aligned}$$

Consequently $n,m \to \infty$ gives $\mathcal{G}(Tr_n,Tr_m,Tr_m) \to 0$. So $\{Tr_n\}$ is $G-$ Cauchy and we have ϖ in X for which $\{Tr_n\}$ converges to ϖ. If T is subsequentially convergent, $\{r_n\}$ has a convergent subsequence $\{r_{n_i}\}$ and

$$\lim_{i\to\infty} r_{n_i} = u \tag{6.34}$$

for a u in X. By continuity of T, $\lim_{i\to\infty} Tr_{n_i} = Tu$. As the limit point is unique, we get

$$Tu = \varpi$$

Whenever f is continuous, it gives $Tfu = Tu$. The injective of T gives $fu = u$. Assume in the case where f is discontinuous, by argument $fr_{n_i} \preceq u$ for all $n_i \in \mathbb{N}$; that is, $r_{n_i+1} \preceq u$. And (6.32) gives

$$\mathcal{G}(Tr_{n_i+1},Tr_{n_i+2},Tfu)$$

$$
\begin{aligned}
&= \mathcal{G}\left(Tfr_{n_i}, Tfr_{n_i+1}, Tfu\right) \\
&\leq \alpha \mathcal{G}\left(Tr_{n_i}, Tr_{n_i+1}, Tu\right) + \beta[\mathcal{G}\left(Tr_{n_i}, Tfr_{n_i}, Tfr_{n_i}\right) + \mathcal{G}\left(Tr_{n_i+1}, Tfr_{n_i+1}, Tfr_{n_i+1}\right) \\
&\quad + \mathcal{G}(Tu, Tfu, Tfu)] + \gamma[\mathcal{G}\left(Tfr_{n_i}, Tfr_{n_i}, Tr_{n_i+1}\right) + \mathcal{G}\left(Tfr_{n_i+1}, Tfr_{n_i+1}, Tr_{n_i}\right) \\
&\quad + \mathcal{G}\left(Tfu, Tfu, Tr_{n_i}\right)] \\
&= \alpha \mathcal{G}\left(Tr_{n_i}, Tr_{n_i+1}, Tu\right) + \beta[\mathcal{G}\left(Tr_{n_i}, Tr_{n_i+1}, Tr_{n_i+1}\right) + \mathcal{G}\left(Tr_{n_i+1}, Tr_{n_i+2}, Tr_{n_i+2}\right) \\
&\quad + \mathcal{G}(Tu, Tfu, Tfu)] + \gamma[\mathcal{G}\left(Tr_{n_i+1}, Tr_{n_i+1}, Tr_{n_i+1}\right) + \mathcal{G}\left(Tr_{n_i+2}, Tr_{n_i+2}, Tr_{n_i}\right) \\
&\quad + \mathcal{G}\left(Tfu, Tfu, Tr_{n_i}\right)]
\end{aligned}
$$

By limiting $i \to \infty$,

$$
\begin{aligned}
\mathcal{G}(Tu, Tu, Tfu) &\leq \alpha \mathcal{G}(Tu, Tu, Tu) + \beta[\mathcal{G}(Tu, Tu, Tu) \\
&\quad + \mathcal{G}(Tu, Tu, Tu) + \mathcal{G}(Tu, Tfu, Tfu)] + \gamma[\mathcal{G}(Tu, Tu, Tu) \\
&\quad + \mathcal{G}(Tu, Tu, Tu) + \mathcal{G}(Tfu, Tfu, Tu)] \\
&= (\beta + \gamma)\mathcal{G}(Tfu, Tu, Tfu) \\
&\leq 2(\beta + \gamma)\mathcal{G}(Tfu, Tu, Tu)
\end{aligned}
$$

gives $\mathcal{G}(Tu, Tu, Tfu) = 0$; that is, $Tu = Tfu$. The injective of T gives $fu = u$.

Now, in case $F(f)$ is well ordered, then we assume that μ is also a fixed point of f. Since u and μ are comparable and $r_{n_i} \preceq u$, by (6.32),

$$
\begin{aligned}
&\mathcal{G}\left(Tr_{n_i+1}, Tu, T\mu\right) \\
&= \mathcal{G}\left(Tfr_{n_i}, Tfu, Tf\mu\right) \\
&\leq \alpha \mathcal{G}\left(Tr_{n_i}, Tu, T\mu\right) + \beta\left[\mathcal{G}\left(Tr_{n_i}, Tfr_{n_i}, Tfr_{n_i}\right) + \mathcal{G}(Tu, Tfu, Tfu) + \mathcal{G}(T\mu, Tf\mu, Tf\mu)\right] \\
&\quad + \gamma\left[\mathcal{G}\left(Tfr_{n_i}, Tfr_{n_i}, Tu\right) + \mathcal{G}\left(Tfu, Tfu, Tr_{n_i}\right) + \mathcal{G}\left(Tr_{n_i}, Tf\mu, Tf\mu\right)\right] \\
&= \alpha \mathcal{G}\left(Tr_{n_i}, Tu, T\mu\right) + \beta\left[\mathcal{G}\left(Tr_{n_i}, Tr_{n_i+1}, Tr_{n_i+1}\right) + \mathcal{G}(Tu, Tu, Tu) + \mathcal{G}(T\mu, T\mu, T\mu)\right] \\
&\quad + \gamma\left[\mathcal{G}\left(Tr_{n_i+1}, Tr_{n_i+1}, Tu\right) + \mathcal{G}\left(Tu, Tu, Tr_{n_i}\right) + \mathcal{G}\left(T\mu, Tr_{n_i}, T\mu\right)\right]
\end{aligned}
$$

$i \to \infty$ implies

$$
\begin{aligned}
\mathcal{G}(Tu, Tu, T\mu) &\leq \alpha \mathcal{G}(Tu, Tu, T\mu) + \beta \mathcal{G}(Tu, Tu, Tu) \\
&\quad + \gamma\left[\mathcal{G}(Tu, Tu, Tu) + \mathcal{G}(Tu, Tu, Tu) + \mathcal{G}(T\mu, Tu, T\mu)\right]
\end{aligned}
$$

$$\leq \alpha\mathcal{G}(Tu,Tu,T\mu)+2\gamma\mathcal{G}(T\mu,Tu,Tu)$$
$$= (\alpha+2\gamma)\mathcal{G}(T\mu,Tu,Tu)$$

which gives $\mathcal{G}(Tu,Tu,T\mu)=0$, and by injectivity of T, we get $u=\mu$.

Corollary 6.8. *In a complete ordered $G-$ metric space $(X,\preceq,\mathcal{G})$, if $T:X\to X$ is an injective and continuous map and $f:X\to X$ is nondecreasing and for all comparable $r,s,u\in X$,*

$$\begin{aligned}\mathcal{G}\left(Tf^m r,Tf^m s,Tf^m u\right) \leq\ & \alpha\mathcal{G}(Tr,Ts,Tu) \\ & +\beta[\mathcal{G}\left(Tr,Tf^m r,Tf^m r\right)+\mathcal{G}\left(Ts,Tf^m s,Tf^m s\right) \\ & +\mathcal{G}\left(Tu,Tf^m u,Tf^m u\right)]+\gamma[\mathcal{G}\left(Tf^m r,Tf^m r,Ts\right) \\ & +\mathcal{G}\left(Tf^m s,Tf^m s,Tr\right)+\mathcal{G}\left(Tf^m u,Tf^m u,Tr\right)]\end{aligned} \tag{6.35}$$

holds for $\alpha,\beta,\gamma\geq 0$ with $\alpha+3\beta+3\gamma<1$. If there is an $r_0\in X$ with $fr_0\succeq r_0$ and either f is G-continuous or a nondecreasing sequence $\{r_n\}$ with $r_n\to z$ in X gives $z_n\preceq z$ for all $n\in\mathbb{N}$, then set $F(f)$ is non-empty provided that T is subsequentially convergent. Moreover, $F(f)$ has at most one element if and only if $F(f)$ is well ordered.

Proof. By Theorem 6.4, it follows $F\left(f^m\right)=\{\varpi\}$ for some $\varpi\in X$. And $f^m\left(f(\varpi)\right)=f^{m+1}(\varpi)=f\left(f^m(\varpi)\right)=f(\varpi)$ gives that $f(\varpi)\in F\left(f^m\right)$. So, we get $\varpi=f(\varpi)$.

Corollary 6.9. *In a complete ordered $G-$ metric space $(X,\preceq,\mathcal{G})$, if $f:X\to X$ is nondecreasing and for all comparable $r,s,u\in X$,*

$$\begin{aligned}\mathcal{G}(fr,fs,fu)\leq \alpha\mathcal{G}(r,s,u)&+\beta\left[\mathcal{G}(r,fr,fr)+\mathcal{G}(s,fs,fs)+\mathcal{G}(u,fu,fu)\right] \\ &+\gamma\left[\mathcal{G}(fr,fr,s)+\mathcal{G}(fs,fs,r)+\mathcal{G}(fu,fu,r)\right]\end{aligned} \tag{6.36}$$

holds for $\alpha,\beta,\gamma\geq 0$ with $\alpha+3\beta+3\gamma<1$. If there is an $r_0\in X$ with $fr_0\succeq r_0$ and either f is G-continuous or a nondecreasing sequence $\{r_n\}$ with $r_n\to z$ in $r_n\to z$ gives $z_n\preceq z$ for all $n\in\mathbb{N}$, then set $F(f)$ is non-empty. Moreover, $F(f)$ has at most one element if and only if $F(f)$ is well ordered.

Proof. We take $T=$ identity map, and by Theorem 6.4, we get the result.

Theorem 6.5. *In a complete ordered $G-$ metric space $(X,\preceq,\mathcal{G})$, if $T:X\to X$ is injective and continuous map and $f:X\to X$ is nondecreasing and for all comparable $r,s,u\in X$,*

$$\mathcal{G}(Tfr,Tfs,Tfu)\leq\lambda\left[\mathcal{G}(Tr,Tfs,Tfs)+\mathcal{G}(Ts,Tfu,Tfu)+\mathcal{G}(Tu,Tfr,Tfr)\right] \tag{6.37}$$

holds for $\lambda \in [0,1)$. *If there is an* $r_0 \in X$ *with* $fr_0 \succeq r_0$ *and either* f *is* G*-continuous or a nondecreasing sequence* $\{r_n\}$ *with* $r_n \to z$ *in* X *gives* $z_n \preceq z$ *for all* $n \in \mathbb{N}$,v *then set* $F(f)$ *is non-empty provided that* T *is subsequentially convergent. Moreover,* $F(f)$ *has at most one element if and only if* $F(f)$ *is well ordered.*

Proof. By nondecreasing of f,

$$r_1 = fr_0 \preceq f^2 r_0 \preceq \ldots \preceq f^n r_0 \preceq f^{n+1} r_0 \preceq \ldots$$

Define $\{r_n\}$ in X with $r_{n+1} = f^n r_0$; that is, $r_{n+1} = fr_{n+1}$ for $n \geq 0$. We assume that $r_n \neq fr_n$ for $n \in \mathbb{N}$; otherwise $r_n = f(r_n)$. Since $r_n \preceq r_{n+1}$ for all $n \geq 0$, by (6.37),

$$\begin{aligned}
& \mathcal{G}(Tr_{n+1}, Tr_{n+2}, Tr_{n+3}) \\
= \ & \mathcal{G}(Tfr_n, Tfr_{n+1}, Tfr_{n+2}) \\
\leq \ & \lambda\left[\mathcal{G}(Tr_n, Tfr_{n+1}, Tfr_{n+1}) + \mathcal{G}(Tr_{n+1}, Tfr_{n+2}, Tfr_{n+2}) + \mathcal{G}(Tr_{n+2}, Tfr_n, Tfr_n)\right] \\
= \ & \lambda\left[\mathcal{G}(Tr_n, Tr_{n+2}, Tr_{n+2}) + \mathcal{G}(Tr_{n+1}, Tr_{n+3}, Tr_{n+3}) + \mathcal{G}(Tr_{n+2}, Tr_{n+1}, Tr_{n+1})\right] \\
\leq \ & \lambda\left[\mathcal{G}(Tr_n, Tr_{n+1}, Tr_{n+2}) + \mathcal{G}(Tr_{n+1}, Tr_{n+2}, Tr_{n+3}) + \mathcal{G}(Tr_{n+1}, Tr_{n+2}, Tr_{n+3})\right] \\
= \ & \lambda\left[\mathcal{G}(Tr_n, Tr_{n+1}, Tr_{n+2}) + 2\mathcal{G}(Tr_{n+1}, Tr_{n+2}, Tr_{n+3})\right]
\end{aligned}$$

that is,

$$(1-2\lambda)\mathcal{G}(Tr_{n+1}, Tr_{n+2}, Tr_{n+3}) \leq \lambda \mathcal{G}(Tr_n, Tr_{n+1}, Tr_{n+2})$$

Hence,

$$\mathcal{G}(Tr_{n+1}, Tr_{n+2}, Tr_{n+3}) \leq k\mathcal{G}(Tr_n, Tr_{n+1}, Tr_{n+2}) \tag{6.38}$$

where $k = \dfrac{\lambda}{1-2\lambda} < 1$. Therefore, for $n \geq 0$,

$$\mathcal{G}(Tr_{n+1}, Tr_{n+2}, Tr_{n+3}) \leq k^{n+1}\mathcal{G}(Tr_0, Tr_1, Tr_2)$$

Following the steps of Theorem 6.2, we find $\{Tr_n\}$ is a $G-$ Cauchy sequence. We have ϖ in X for which $\{Tr_n\}$ converges to ϖ. If T is subsequentially convergent, $\{r_n\}$ has a convergent subsequence $\{r_{n_i}\}$ and

$$\lim_{i \to \infty} r_{n_i} = u \tag{6.39}$$

for a u in X. By continuity of T, $\lim\limits_{i \to \infty} Tr_{n_i} = Tu$. As the limit point is unique, we get

$$Tu = \varpi$$

Whenever f is continuous, it gives $Tfu = Tu$. The injective of T gives $fu = u$. In the case where f is discontinuous, by argument $fr_{n_i} \preceq u$ for all $n_i \in \mathbb{N}$; that is, $r_{n_i+1} \preceq u$. And with (6.37), we have

$$\begin{aligned}
&\mathcal{G}\left(Tr_{n_i+1}, Tr_{n_i+2}, Tfu\right)\\
&= \mathcal{G}\left(Tfr_{n_i}, Tfr_{n_i+1}, Tfu\right)\\
&\leq \lambda\left[\mathcal{G}\left(Tr_{n_i}, Tfr_{n_i+1}, Tfr_{n_i+1}\right) + \mathcal{G}\left(Tr_{n_i+1}, Tfu, Tfu\right) + \mathcal{G}\left(Tu, Tfr_{n_i}, Tfr_{n_i}\right)\right]\\
&= \lambda\left[\mathcal{G}\left(Tr_{n_i}, Tr_{n_i+2}, Tr_{n_i+2}\right) + \mathcal{G}\left(Tr_{n_i+1}, Tfu, Tfu\right) + \mathcal{G}\left(Tu, Tr_{n_i+1}, Tr_{n_i+1}\right)\right]
\end{aligned}$$

As $i \to \infty$ gives

$$\begin{aligned}
\mathcal{G}(Tfu, Tu, Tu) &\leq \lambda\left[\mathcal{G}(Tu, Tu, Tu) + \mathcal{G}(Tu, Tfu, Tfu) + \mathcal{G}(Tu, Tu, Tu)\right]\\
&= \lambda\mathcal{G}(Tu, Tfu, Tfu)\\
&\leq 2\lambda\mathcal{G}(Tu, Tu, Tfu)
\end{aligned}$$

gives $\mathcal{G}(Tu, Tu, Tfu) = 0$; that is, $Tu = Tfu$. The injective of T gives $fu = u$.

Now, $F(f)$ is well ordered. Then we assume that v is also a fixed point of f. Since u and v are comparable and $r_{n_i} \preceq u$, by (6.37),

$$\begin{aligned}
&\mathcal{G}\left(Tr_{n_i+1}, Tu, Tv\right)\\
&= \mathcal{G}\left(Tfr_{n_i}, Tfu, Tfv\right)\\
&\leq \lambda\left[\mathcal{G}\left(Tr_{n_i}, Tfu, Tfu\right) + \mathcal{G}(Tu, Tfv, Tfv) + \mathcal{G}\left(Tv, Tfr_{n_i}, Tfr_{n_i}\right)\right]\\
&= \lambda\left[\mathcal{G}\left(Tr_{n_i}, Tu, Tu\right) + \mathcal{G}(Tu, Tv, Tv) + \mathcal{G}\left(Tv, Tr_{n_i+1}, Tr_{n_i+1}\right)\right]
\end{aligned}$$

And by $i \to \infty$,

$$\begin{aligned}
\mathcal{G}(Tu, Tu, Tv) &\leq \lambda\left[\mathcal{G}(Tu, Tv, Tv) + \mathcal{G}(Tv, Tu, Tu)\right]\\
&= \lambda\left[\mathcal{G}(Tu, Tv, Tv) + \mathcal{G}(Tv, Tu, Tu)\right]\\
&\leq 3\lambda\mathcal{G}(Tu, Tv, Tv)
\end{aligned}$$

which gives $\mathcal{G}(Tu, Tu, Tv) = 0$, and by injectivity of T, we get $u = v$.

Example 6.6. *Let $X = \{0,1,2\}$ be a set with usual order. Let $\mathcal{G} : X \times X \times X \to [0,\infty)$ be given by*

(r,u,s)	$\mathcal{G}(r,u,s)$
$(0,0,0),(1,1,1),(2,2,2),$	0
$(0,1,1),(1,0,1),(1,1,0),$	1
$(1,0,0),(0,1,0),(0,0,1),(1,2,2),(2,1,2),(2,2,1),$	2
$(0,0,2),(0,2,0),(2,0,0),(0,2,2),(2,0,2),(2,2,0),$	3
$(1,1,2),(1,2,1),(2,1,1),(0,1,2),(0,2,1),(1,0,2),(1,2,0),(2,0,1),(2,1,0),$	4.

As $\mathcal{G}(0,1,0) \neq \mathcal{G}(1,0,1)$; that is, $\mathcal{G}$ is non-symmetric. Self-maps T and f are defined on X as $\to X$

r	$T(r)$	$f(r)$
0	1	0
1	0	0
2	2	1

Now, we check condition (6.37) for $r,s,u \in X$:

(1) $r = 0,\ s = 1,\ u = 2$

$$\begin{aligned}\mathcal{G}(Tfr,Tfs,Tfu) = G(1,1,0) = 1 \le \lambda[0+0+4]\\ &= \lambda[\mathcal{G}(1,1,1)+\mathcal{G}(0,0,0)+\mathcal{G}(2,1,1)]\\ &= \lambda[\mathcal{G}(Tr,Tfs,Tfs)+\mathcal{G}(Ts,Tfu,Tfu)+\mathcal{G}(Tu,Tfr,Tfr)]\end{aligned}$$

(2) $r = 0,\ s = 2,\ u = 1$

$$\begin{aligned}\mathcal{G}(Tfr,Tfs,Tfu) &= \mathcal{G}(1,0,1) = 1 \le \lambda[2+4+1]\\ &= \lambda[\mathcal{G}(1,0,0)+\mathcal{G}(2,1,1)+\mathcal{G}(0,1,1)]\\ &= \lambda[\mathcal{G}(Tr,Tfs,Tfs)+\mathcal{G}(Ts,Tfu,Tfu)+\mathcal{G}(Tu,Tfr,Tfr)]\end{aligned}$$

(3) $r = 1,\ s = 0,\ u = 2$

$$\begin{aligned}\mathcal{G}(Tfr,Tfs,Tfu) &= \mathcal{G}(1,1,0) = 1 \le \lambda[1+2+4]\\ &= \lambda[\mathcal{G}(0,1,1)+\mathcal{G}(1,0,0)+\mathcal{G}(2,1,1)]\\ &= \lambda[\mathcal{G}(Tr,Tfs,Tfs)+\mathcal{G}(Ts,Tfu,Tfu)+\mathcal{G}(Tu,Tfr,Tfr)]\end{aligned}$$

(4) $r=1,\ s=2,\ u=0$

$$\begin{aligned}\mathcal{G}(Tfr,Tfs,Tfu) &= \mathcal{G}(1,0,1)=1\le\lambda[0+4+0]\\ &= \lambda[\mathcal{G}(0,0,0)+\mathcal{G}(2,1,1)+\mathcal{G}(1,1,1)]\\ &= \lambda[\mathcal{G}(Tr,Tfs,Tfs)+\mathcal{G}(Ts,Tfu,Tfu)+\mathcal{G}(Tu,Tfr,Tfr)]\end{aligned}$$

(5) $r=2,\ s=0,\ u=1$

$$\begin{aligned}\mathcal{G}(Tfr,Tfs,Tfu) &= \mathcal{G}(0,1,1)=1\le\lambda[4+0+0]\\ &= \lambda[\mathcal{G}(2,1,1)+\mathcal{G}(1,1,1)+\mathcal{G}(0,0,0)]\\ &= \lambda[\mathcal{G}(Tr,Tfs,Tfs)+\mathcal{G}(Ts,Tfu,Tfu)+\mathcal{G}(Tu,Tfr,Tfr)]\end{aligned}$$

(6) $r=2,\ s=1,\ u=0$

$$\begin{aligned}\mathcal{G}(Tfr,Tfs,Tfu) &= \mathcal{G}(0,1,1)=1\le\lambda[4+1+2]\\ &= \lambda[\mathcal{G}(2,1,1)+\mathcal{G}(0,1,1)+\mathcal{G}(1,0,0)]\\ &= \lambda[\mathcal{G}(Tr,Tfs,Tfs)+\mathcal{G}(Ts,Tfu,Tfu)+\mathcal{G}(Tu,Tfr,Tfr)]\end{aligned}$$

Notice that condition (6.41) holds for $\lambda=\frac{7}{22}$, *as all conditions of Theorem 6.5 hold. Moreover,* $F(f)=\{0\}$.

Corollary 6.10. *In a complete ordered* $G-$ *metric space* $(X,\preceq,\mathcal{G})$, *if* $T:X\to X$ *is injective and continuous map and* $f:X\to X$ *is nondecreasing and for all comparable* $r,s,u\in X$,

$$\mathcal{G}\left(Tf^m r,Tf^m s,Tf^m u\right) \le \lambda[\mathcal{G}\left(Tr,Tf^m s,Tf^m u\right)+\mathcal{G}\left(Ts,Tf^m u,Tf^m u\right) + \mathcal{G}\left(Tu,Tf^m r,Tf^m r\right)] \tag{6.40}$$

holds for $\lambda\in[0,1)$. *If there is an* $r_0\in X$ *with* $fr_0\succeq r_0$ *and either* f *is G-continuous or a nondecreasing sequence* $\{r_n\}$ *with* $r_n\to z$ *in* X *gives* $z_n\preceq z$ *for all* $n\in\mathbb{N}$, *then set* $F(f)$ *is non-empty provided that* T *is subsequentially convergent. Moreover,* $F(f)$ *has at most one element if and only if* $F(f)$ *is well ordered.*

Now, if we take T as an identity map in Theorem 6.5, then the following corollary follows.

Corollary 6.11. *In a complete ordered* $G-$ *metric space* $(X,\preceq,\mathcal{G})$, *if* $f:X\to X$ *is nondecreasing and for all comparable* $r,s,u\in X$,

$$\mathcal{G}(fr,fs,fu)\le\lambda[\mathcal{G}(r,fs,fu)+\mathcal{G}(s,fu,fu)+\mathcal{G}(u,fr,fr)] \tag{6.41}$$

holds for $\lambda \in [0,1)$. *If there is an* $r_0 \in X$ *with* $fr_0 \succeq r_0$ *and either* f *is* G*-continuous or a nondecreasing sequence* $\{r_n\}$ *havwithing* $r_n \to z$ *in* X *gives* $z_n \preceq z$ *for all* $n \in \mathbb{N}$, *then set* $F(f)$ *is non-empty. Moreover,* $F(f)$ *has at most one element if and only if* $F(f)$ *is well ordered.*

Remark 6.1. *The conditions (6.32), (6.35), (6.36), (6.37), (6.40) and (6.41) cannot convert in the setting of metric* d_G. *Therefore, the previous result cannot deduce to* $(X, d_{\mathcal{G}})$.

6.6 PERIODIC POINT PROPERTY OF SELF-MAPS IN G – METRIC SPACES

For a self-map T, if we have $T(p) = p$, then we also get $T^m(p) = p$ for every natural number m. But the converse is not true. For example, consider $X = [0,1]$, and T defined on X by $Tx = 1 - x$. Note that $T\left(\frac{1}{2}\right) = \frac{1}{2}$ and T^2 is an identity map, all of whose points in $[0,1]$ are its fixed point. Also, when $= [0,\pi]$, $Tx = \cos x, T$, and each of its iterations has the same fixed point [27–29]. When $F(T) = F(T^l)$ for each $l \in \mathbb{N}$ then we say that T has property P. Rhoades and Jeong [28] established many results of contractions that have the property P. Abbas et al. [30] and Chugh et al. [31] studied the property P in $G-$ metric space. In this section, first we obtain the periodic point results in $G-$ metric space.

Theorem 6.6. *In a complete* $G-$ *metric space* $(X, \mathcal{G})$, *if* $f : X \to X$ *is continuous and satisfies*

$$\mathcal{G}(fv, fv, f^2v) \le h\mathcal{G}(v, v, fv) \tag{6.42}$$

or

$$\mathcal{G}(fv, f^2v, f^2v) \le h\mathcal{G}(v, fv, fv) \tag{6.43}$$

for all $v \in X$ *having* $0 \le h < 1$, *then* $F(f) \ne \phi$ *and* f *has a property* P.

Proof. First, we show that $F(f) \ne \phi$. In the case of a symmetric $G-$ metric, by adding (6.42) and (6.43), we get

$$\mathcal{G}(fv, fv, f^2v) + \mathcal{G}(fv, f^2v, f^2v)$$
$$\le h[G(v, v, fv) + G(v, fv, fv)]$$

which gives

$$d_{\mathcal{G}}(fv, f^2v) \le hd_{\mathcal{G}}(v, fv) \tag{6.44}$$

for all $\upsilon \in X$ with $0 \leq h < 1$, and $F(f)$ is non-empty, which follows from [32].

For a non-symmetric $F(f)$ metric, we get

$$\begin{aligned} d_{\mathcal{G}}\left(f\upsilon, f^2\upsilon\right) &= \mathcal{G}\left(f\upsilon, f\upsilon, f^2\upsilon\right) + \mathcal{G}\left(f\upsilon, f^2\upsilon, f^2\upsilon\right) \\ &\leq h\left[\mathcal{G}(\upsilon,\upsilon,f\upsilon) + \mathcal{G}(\upsilon, f\upsilon, f\upsilon)\right] \\ &\leq h\left[\frac{2}{3}d_{\mathcal{G}}(\upsilon, f\upsilon) + \frac{2}{3}d_{\mathcal{G}}(\upsilon, f\upsilon)\right] \\ &= \frac{4h}{3}d_{\mathcal{G}}(\upsilon, f\upsilon) \end{aligned}$$

for all $\upsilon \in X$. Here, the contractivity factor $\frac{4h}{3}$ can be greater than 1, so there is no information in the metric space. So for $u \in X$, we take a sequence $\{r_n\}$ with $r_n = f^n(u)$, Therefore, from (6.42),

$$\begin{aligned} \mathcal{G}(r_{n+1}, r_{n+1}, r_n) &\leq h\mathcal{G}(r_n, r_{n-1}, r_{n-1}) \\ &\leq h^2\mathcal{G}(r_{n-1}, r_{n-1}, r_{n-2}) \\ &\leq \ldots \leq h^n\mathcal{G}(u,u,fu) \end{aligned}$$

Now, for any m, n where $n < m$,

$$\begin{aligned} \mathcal{G}(r_n, r_n, r_m) &\leq \mathcal{G}(r_n, r_n, r_{n+1}) + \mathcal{G}(r_{n+1}, r_{n+1}, r_{n+2}) \\ &\quad + \ldots + \mathcal{G}(r_{m-1}, r_{m-1}, r_m) \\ &\leq \left[h^n + h^{n+1} + \ldots + h^m\right]\mathcal{G}(fu,u,u) \\ &\leq \frac{h^m}{1-h}\mathcal{G}(fu,u,u) \end{aligned}$$

and so $\mathcal{G}(r_n, r_n, r_m) \to 0$ as $m, n \to \infty$. We obtain that $\{r_n\}$ is a $G-$Cauchy sequence. There is $\varpi \in X$ for which $\{r_n\}$ $G-$converges to ϖ. Now

$$\begin{aligned} \mathcal{G}\left(f^n u, f^n u, f^{n+1}u\right) &\leq h\mathcal{G}\left(f^{n-1}u, f^{n-1}u, f^n u\right) \\ &\leq h^2\mathcal{G}\left(f^{n-2}u, f^{n-2}u, f^{n-1}u\right) \\ &\leq \ldots \leq h^n\mathcal{G}(u,u,fu) \end{aligned}$$

Limiting $n \to \infty$ gives

$$\mathcal{G}(\varpi, \varpi, f\varpi) \leq 0$$

that is, $\varpi = f\varpi$. Therefore, $F(f) \neq \phi$.

We shall always consider $n > 1$, as for $n = 1$, it is true, obviously. Consider $\upsilon \in F(f^n)$. Then

$$\begin{aligned} \mathcal{G}(\upsilon,\upsilon,f\upsilon) &= \mathcal{G}\left(f\left(f^{n-1}\upsilon\right), f\left(f^{n-1}\upsilon\right), f^2\left(f^{n-1}\upsilon\right)\right) \\ &\leq h\mathcal{G}\left(f^{n-1}\upsilon, f^{n-1}\upsilon, f^n\upsilon\right) \\ &= h\mathcal{G}\left(f\left(f^{n-2}\upsilon\right), f\left(f^{n-2}\upsilon\right), f^2\left(f^{n-2}\upsilon\right)\right) \\ &\leq h^2\mathcal{G}\left(f^{n-2}\upsilon, f^{n-2}\upsilon, f^{n-1}\upsilon\right) \\ &\leq \ldots \leq h^n\mathcal{G}(\upsilon,\upsilon,f\upsilon) \end{aligned}$$

As $n > 1$ is fixed and $0 \leq h < 1$, so $\mathcal{G}(\upsilon,\upsilon,f\upsilon) \leq h\mathcal{G}(f\upsilon,\upsilon,\upsilon)$; that is, $\upsilon = f\upsilon$, and hence $F(f^n) = F(f)$.

Remark 6.2. *In Theorem 6.6, if equations (6.42) and (6.43) are satisfied with* $0 \leq h < \frac{3}{4}$, *then whenever we have a non-symmetric* $G-$ *metric space, we obtain*

$$d_{\mathcal{G}}\left(fr, f^2r\right) \leq \frac{4h}{3} d_{\mathcal{G}}(r, fr)$$

where $\frac{4h}{3} < 1$, *and the fixed point of* f *follows from [32].*

Theorem 6.7. *In a complete* $G-$ *metric space* (X,G), *if* $f : X \to X$ *is continuous and satisfies*

$$G(fr, f^2r, f^3r) \leq hG(r, fr, f^2r) \tag{6.45}$$

for all $r \in X$ *with* $0 \leq h < 1$, *then* $F(f) \neq \phi$ *and* f *has property* P.

Proof. For $u \in X$, consider a sequence $\{r_n\}$ by $r_n = f^n(u)$, For $n \geq 0$,

$$\begin{aligned} G(r_n, r_{n+1}, r_{n+2}) &\leq hG(r_{n-1}, r_n, r_{n+1}) \\ &\leq h^2G(r_{n-2}, r_{n-1}, r_n) \\ &\leq \ldots \leq h^n\mathcal{G}\left(u, fu, f^2u\right) \end{aligned}$$

Now, for m,l,n with $n < m < l$,

$$\begin{aligned} G(r_n, r_m, r_l) \leq\ & G(r_n, r_{n+1}, r_{n+1}) + G(r_{n+1}, r_{n+1}, r_{n+2}) \\ & + \cdots + \mathcal{G}(r_{l-1}, r_{l-1}, r_l) \\ \leq\ & \mathcal{G}(r_{n+2}, r_{n+1}, r_n) + \mathcal{G}(r_{n+2}, r_{n+1}, r_n) \end{aligned}$$

$$
\begin{aligned}
&+\ldots+\mathcal{G}(r_{l-2},r_{l-1},r_l)\\
&\le \left[h^n+h^{n+1}+\ldots+h^l\right]\mathcal{G}\left(f^2u,fu,u\right)\\
&\le \frac{h^l}{1-h}\mathcal{G}\left(f^2u,fu,u\right)
\end{aligned}
$$

and so $\mathcal{G}(r_n,r_n,r_m)\to 0$ as $m,n\to\infty$. So $\{r_n\}$ is $G-$ Cauchy sequence. So there is $\varpi\in X$ for which $\{r_n\}$ $G-$ converges to ϖ. Now by (6.45),

$$
\begin{aligned}
\mathcal{G}\left(f^nu,f^{n+1}u,f^{n+2}u\right) &\le h\mathcal{G}\left(f^{n-1}u,f^nu,f^{n+1}u\right)\\
&\le h^2\mathcal{G}\left(f^{n-2}u,f^{n-1}u,f^{n-2}u\right)\\
&\le \ldots\le h^n\mathcal{G}\left(u,fu,f^2u\right)
\end{aligned}
$$

And by $n\to\infty$,

$$
\mathcal{G}\left(\varpi,f\varpi,f^2\varpi\right)\le 0
$$

and $\varpi=f\varpi$. So $F(f)\neq\phi$.

Now consider $u\in F\left(f^n\right)$; then

$$
\begin{aligned}
\mathcal{G}\left(u,fu,f^2u\right) &= \mathcal{G}\left(f\left(f^{n-1}u\right),f^2\left(f^{n-1}u\right),f^3\left(f^{n-1}u\right)\right)\\
&\le h\mathcal{G}\left(f^{n-1}u,f^nu,f^{n+1}u\right)\\
&= h\mathcal{G}\left(f\left(f^{n-2}u\right),f^2\left(f^{n-2}u\right),f^3\left(f^{n-2}u\right)\right)\\
&\le h^2\mathcal{G}\left(f^nu,f^{n-1}u,f^{n-2}u\right)\\
&\le \ldots\le h^n\mathcal{G}\left(f^2u,fu,u\right)
\end{aligned}
$$

This gives $\mathcal{G}\left(f^2u,fu,u\right)\le h\mathcal{G}\left(f^2u,fu,u\right)$, and so $fu=u$ and hence $F\left(f^n\right)=F(f)$.

Example 6.7. Let $X=[0,1]$ *and* $\mathcal{G}(r,s,r)=max\{|r-s|,|s-r|,|r-r|\}$ *be a* $G-$*metric on* X. *A continuous map* $f:X\to X$ *is defined as*

$$
fr=\frac{\alpha}{\beta}r
$$

where $0<\alpha<\beta$. *Consider*

$$\mathcal{G}\left(fr, f^2r, f^3r\right) = \max\left(\left|\frac{\alpha}{\beta}r - \frac{\alpha^2}{\beta^2}r\right|, \left|\frac{\alpha^2}{\beta^2}r - \frac{\alpha^3}{\beta^3}r\right|, \left|\frac{\alpha^3}{\beta^3}r - \frac{\alpha}{\beta}r\right|\right)$$

$$= \left(\frac{\alpha}{\beta}r - \frac{\alpha^3}{\beta^3}r\right)$$

$$= \frac{\alpha}{\beta}\left(r - \frac{\alpha^2}{\beta^2}r\right)$$

$$\leq \frac{\alpha}{\beta}\max\left(\left|r - \frac{\alpha}{\beta}r\right|, \left|\frac{\alpha}{\beta}r - \frac{\alpha^2}{\beta^2}r\right|, \left|r - \frac{\alpha^2}{\beta^2}r\right|\right)$$

$$= h\mathcal{G}\left(r, fr, f^2r\right)$$

Hence (6.45) is satisfied for $h = \frac{\alpha}{\beta}$. *Obviously* $\left\{f^n r\right\}$ *converges to* 0 *where* $0 = f(0)$ *and* $F(f) = F\left(f^n\right)$.

Corollary 6.12. *In a complete* G*–metric space* $(X, \mathcal{G})$, *if* $f : X \to X$ *is continuous and satisfies*

$$\begin{aligned}\mathcal{G}(fr, fs, fu) \leq\ & \text{kmax}\{\mathcal{G}(r,s,u), \mathcal{G}(r, fr, fs)\mathcal{G}(s, fs, fu)/\mathcal{G}(r,s,u) \\ & \mathcal{G}(fr, fr, s)\mathcal{G}(fs, fs, u)/\mathcal{G}(r,s,u) \\ & \mathcal{G}(r, fr, u)\mathcal{G}(s, fs, u)/2\mathcal{G}(r,s,u)\}\end{aligned} \tag{6.46}$$

for all $r, s, u \in X$ *with* $0 \leq k < 1$, *t hen* $F(f) \neq \phi$ *and* f *has property P.*

Proof. Taking $s = fr,\ u = f^2r$ in (6.46), we obtain

$$\begin{aligned}&\mathcal{G}\left(fr, f^2r, f^3r\right) \\ \leq\ & \text{kmax}\{\mathcal{G}\left(r, fr, f^2r\right), \mathcal{G}\left(r, fr, f^2r\right)\mathcal{G}\left(fr, f^2r, f^3r\right)/\mathcal{G}\left(r, fr, f^2r\right) \\ & \mathcal{G}(fr, fr, fr)\mathcal{G}\left(f^2r, f^2r, f^2r\right)/\mathcal{G}\left(r, fr, fr^2\right) \\ & \mathcal{G}\left(r, fr, f^2r\right)\mathcal{G}\left(fr, f^2r, f^2r\right)/2\mathcal{G}\left(r, fr, f^2r\right)\} \\ =\ & \text{kmax}\left\{\mathcal{G}\left(r, fr, f^2r\right), \mathcal{G}\left(fr, f^2r, f^3r\right), 0, \mathcal{G}\left(fr, f^2r, f^2r\right)/2\right\} \\ =\ & k\mathcal{G}\left(r, fr, f^2r\right)\end{aligned}$$

Hence the result follows from Theorem 6.7.

Corollary 6.13. *In a complete $G-$ metric space $(X,\mathcal{G})$, if $f:X\to X$ is continuous and satisfies*

$$\mathcal{G}(fr,fs,fu)\le h\frac{\mathcal{G}(fr,s,fs)\mathcal{G}(fs,s,fu)}{\mathcal{G}(fr,s,u)+\mathcal{G}(fr,s,s)}+k\mathcal{G}(r,s,u) \tag{6.47}$$

for all $r,s,u\in X$ where $h,k\in[0,1)$ and $h+k<1$, then $F(f)\neq\phi$ and f has property P.

Proof. For $u=fr,\ s=f^2r,$ from (6.47),

$$\mathcal{G}\left(fr,f^2r,f^3r\right)\le h\frac{\mathcal{G}\left(fr,fr,f^3r\right)\mathcal{G}\left(f^2r,fr,f^3r\right)}{\mathcal{G}\left(fr,fr,f^2r\right)+\mathcal{G}(fr,fr,fr)}+k\mathcal{G}\left(r,fr,f^2r\right)$$
$$=h\mathcal{G}\left(fr,f^2r,f^3r\right)+k\mathcal{G}\left(r,fr,f^2r\right)$$

Thus

$$\mathcal{G}\left(fr,f^2r,f^3r\right)\le\frac{k}{1-h}\mathcal{G}\left(r,fr,f^2r\right)$$

where $0\le\frac{k}{1-h}<1$. Thus (6.47) is satisfied, and proof can be obtained by Theorem 6.7.

Theorem 6.8. *In a complete $G-$metric space $(X,\mathcal{G})$, if $f:X\to X$ is continuous and satisfies*

$$\begin{aligned}\mathcal{G}(fw,fs,fu)\le{}& h\max\{\mathcal{G}(w,s,u),\mathcal{G}(w,fw,fw),\mathcal{G}(s,fs,fs),\mathcal{G}(u,fu,fu)\\&[\mathcal{G}(fw,fw,s)+\mathcal{G}(fs,fs,u)+\mathcal{G}(s,fu,fu)]/2\}\\&+k\max\{\mathcal{G}(w,fw,fw),\mathcal{G}(s,fs,fs),\mathcal{G}(u,fu,fu)\}\\&+l[\mathcal{G}(w,fw,fw)+\mathcal{G}(s,fs,fs)+\mathcal{G}(u,fu,fu)]\end{aligned} \tag{6.48}$$

for all $w,s,u\in X$ where $h,k,l\ge0$ and $h+k+l<\frac{1}{3}$. Then $F(f)\neq\phi$ and f has property P.

Proof. It follows from (6.48), by taking $s,w\in X,$ we obtain

$$\begin{aligned}&\mathcal{G}(fw,fs,fs)\\&\le h\max\{\mathcal{G}(w,s,s),\mathcal{G}(w,fw,fw),\mathcal{G}(s,fs,fs),\mathcal{G}(s,fs,fs)\\&[\mathcal{G}(fw,fw,s)+\mathcal{G}(fs,fs,s)+\mathcal{G}(s,fs,fs)]/2\}\end{aligned}$$

$$
\begin{aligned}
&+k\max\{\mathcal{G}(w,fw,fw),\mathcal{G}(s,fs,fs),\mathcal{G}(s,fs,fs)\}\\
&+l[\mathcal{G}(w,fw,fw)+\mathcal{G}(s,fs,fs)+\mathcal{G}(s,fs,fs)]\\
&=h\max\{\mathcal{G}(w,s,s),\mathcal{G}(w,fw,fw),\mathcal{G}(s,fs,fs),[\mathcal{G}(fw,fw,s)+2\mathcal{G}(fs,fs,s)]/2\}\\
&+k\max\{\mathcal{G}(w,fw,fw),\mathcal{G}(s,fs,fs)\}+l[\mathcal{G}(w,fw,fw)+2\mathcal{G}(s,fs,fs)]
\end{aligned}
$$

Now, for any $r_0 \in X$, we take a sequence $\{r_n\}$ by $r_n = f^n(r_0)$. Then for all $m \geq 0$,

$$
\begin{aligned}
&\mathcal{G}(r_m,r_{m+1},r_{m+1})\\
&\leq h\max\{\mathcal{G}(r_{m-1},r_m,r_m),\mathcal{G}(r_{m-1},r_m,r_m),\mathcal{G}(r_m,r_{m+1},r_{m+1})\\
&\quad[\mathcal{G}(r_m,r_m,r_m)+2\mathcal{G}(r_{m+1},r_{m+1},r_m)]/2\}\\
&\quad+k\max\{\mathcal{G}(r_{m-1},r_m,r_m),\mathcal{G}(r_m,r_{m+1},r_{m+1})\}\\
&\quad+l[\mathcal{G}(r_{m-1},r_m,r_m)+2\mathcal{G}(r_m,r_{m+1},r_{m+1})]\\
&\leq (h+k)\max\{\mathcal{G}(r_{m-1},r_m,r_m),\mathcal{G}(r_m,r_{m+1},r_{m+1})\\
&\quad\mathcal{G}(r_{m-1},r_m,r_m)+2\mathcal{G}(r_m,r_{m+1},r_{m+1})\}\\
&= (h+k+l)[\mathcal{G}(r_{m-1},r_m,r_m)+2\mathcal{G}(r_m,r_{m+1},r_{m+1})]
\end{aligned}
$$

implies that

$$
\mathcal{G}(r_m,r_{m+1},r_{m+1}) \leq \lambda\mathcal{G}(r_{m-1},r_m,r_m)
$$

where $\lambda = \dfrac{(h+k+l)}{1-2(h+k+l)}$. Obviously $0 \leq \lambda < 1$. Thus for all $m \geq 0$,

$$
\begin{aligned}
\mathcal{G}(r_m,r_{m+1},r_{m+1}) &\leq \lambda\mathcal{G}(r_{m-1},r_m,r_m)\\
&\leq \lambda^2\mathcal{G}(r_{m-2},r_{m-1},r_{m-1})\\
&\leq \ldots \leq \lambda^m\mathcal{G}(r_0,r_1,r_1)
\end{aligned}
$$

Now, for n,m with $n < m$,

$$
\begin{aligned}
G(r_m,r_m,r_n). &\leq \mathcal{G}(r_m,r_{m-1},r_{m-1})+\mathcal{G}(r_{m-1},r_{m-1},r_{m-2})\\
&\quad+\ldots+\mathcal{G}(r_{n+1},r_{n+1},r_n)\\
&\leq [\lambda^n+\lambda^{n+1}+\ldots+\lambda^m]\mathcal{G}(r_0,r_1,r_1)\\
&\leq \frac{\lambda^m}{1-\lambda}\mathcal{G}(r_0,r_1,r_2)
\end{aligned}
$$

We obtain that $\{r_n\}$ is a $\mathcal{G}-$Cauchy sequence, and there is $\varpi \in X$ for which $\{r_n\}$ $\mathcal{G}-$converges to ϖ. Assume that $f\varpi \neq \varpi$. Then

$$\begin{aligned}
&\mathcal{G}(r_n, f\varpi, f\varpi)\\
&\leq h\max\{\mathcal{G}(r_{n-1},\varpi,\varpi),\mathcal{G}(r_{n-1},r_n,r_n),\mathcal{G}(\varpi,f\varpi,f\varpi),\mathcal{G}(\varpi,f\varpi,f\varpi)\\
&\quad[\mathcal{G}(r_n,r_n,\varpi)+\mathcal{G}(f\varpi,f\varpi,\varpi)+\mathcal{G}(f\varpi,f\varpi,\varpi)]/2\}\\
&\quad+k\max\{\mathcal{G}(r_{n-1},r_n,r_n),\mathcal{G}(\varpi,f\varpi,f\varpi),\mathcal{G}(\varpi,f\varpi,f\varpi)\}\\
&\quad+l[\mathcal{G}(r_{n-1},r_n,r_n)+\mathcal{G}(\varpi,f\varpi,f\varpi)+\mathcal{G}(\varpi,f\varpi,f\varpi)]\\
&= h\max\{\mathcal{G}(r_{n-1},\varpi,\varpi),\mathcal{G}(r_{n-1},r_n,r_n),\mathcal{G}(\varpi,f\varpi,f\varpi)\\
&\quad[\mathcal{G}(r_n,r_n,\varpi)+2\mathcal{G}(f\varpi,f\varpi,\varpi)]/2\}+k\max\{\mathcal{G}(r_{n-1},r_n,r_n),\mathcal{G}(\varpi,f\varpi,f\varpi)\}\\
&\quad+l[\mathcal{G}(r_{n-1},r_n,r_n)+2\mathcal{G}(\varpi,f\varpi,f\varpi)]
\end{aligned}$$

Taking $n \to \infty$ implies

$$\begin{aligned}
\mathcal{G}(\varpi,f\varpi,f\varpi) &\leq h\max\{0,\mathcal{G}(\varpi,f\varpi,f\varpi)\}+k\max\{0,\mathcal{G}(\varpi,f\varpi,f\varpi)\}\\
&\quad+2l\mathcal{G}(\varpi,f\varpi,f\varpi)\\
&=(h+k+2l)\mathcal{G}(\varpi,f\varpi,f\varpi)
\end{aligned}$$

a contradiction. Hence $F(f) \neq \phi$.

For $n > 1$, consider $\varpi \in F(f^n)$. Using (6.48),

$$\begin{aligned}
&\mathcal{G}(\varpi,f\varpi,f\varpi)\\
&= \mathcal{G}\left(f\left(f^{n-1}\varpi\right),f\left(f^n\varpi\right),f\left(f^n\varpi\right)\right)\\
&\leq h\max\{\mathcal{G}\left(f^{n-1}\varpi,f^n\varpi,f^n\varpi\right),\mathcal{G}\left(f^{n-1}\varpi,f^n\varpi,f^n\varpi\right),\mathcal{G}\left(f^n\varpi,f^{n+1}\varpi,f^{n+1}\varpi\right)\\
&\quad\left[\mathcal{G}\left(f^n\varpi,f^n\varpi,f^n\varpi\right)+2\mathcal{G}\left(f^{n+1}\varpi,f^{n+1}\varpi,f^n\varpi\right)\right]/2\}\\
&\quad+k\max\left\{\mathcal{G}\left(f^{n-1}\varpi,f^n\varpi,f^n\varpi\right),\mathcal{G}\left(f^n\varpi,f^{n+1}\varpi,f^{n+1}\varpi\right)\right\}\\
&\quad+l\left[\mathcal{G}\left(f^{n-1}\varpi,f^n\varpi,f^n\varpi\right)+2\mathcal{G}\left(f^n\varpi,f^{n+1}\varpi,f^{n+1}\varpi\right)\right]\\
&= h\max\left\{\mathcal{G}\left(f^{n-1}\varpi,\varpi,\varpi\right),\mathcal{G}(\varpi,f\varpi,f\varpi),[2\mathcal{G}(f\varpi,f\varpi,\varpi)]/2\right\}\\
&\quad+k\max\left\{\mathcal{G}\left(f^{n-1}\varpi,\varpi,\varpi\right),\mathcal{G}(\varpi,f\varpi,f\varpi)\right\}+l\left[\mathcal{G}\left(f^{n-1}\varpi,\varpi,\varpi\right)+2\mathcal{G}(\varpi,f\varpi,f\varpi)\right]\\
&\leq (h+k+l)\left[\mathcal{G}\left(f^{n-1}\varpi,\varpi,\varpi\right)+2\mathcal{G}(\varpi,f\varpi,f\varpi)\right]
\end{aligned}$$

Thus for $\lambda := h+k+l<1$,

$$\mathcal{G}(\varpi, f\varpi, f\varpi) \leq \lambda\left[\mathcal{G}\left(f^{n-1}\varpi, \varpi, \varpi\right) + 2\mathcal{G}(\varpi, f\varpi, f\varpi)\right]$$

which further gives

$$\mathcal{G}(\varpi, f\varpi, f\varpi) \leq \eta\mathcal{G}\left(f^{n-1}\varpi, \varpi, \varpi\right)$$

where $\eta = \dfrac{\lambda}{1-2\lambda}$. Obviously $0 \leq \eta < 1$. Thus for

$$\begin{aligned}
\mathcal{G}(\varpi, f\varpi, f\varpi) &\leq \eta\mathcal{G}\left(f^{n-1}\varpi, \varpi, \varpi\right) \\
&= \eta\mathcal{G}\left(f^{n-1}\varpi, f^{n}\varpi, f^{n}\varpi\right) \\
&\leq \eta^2\mathcal{G}\left(f^{n-2}\varpi, f^{n-1}\varpi, f^{n-1}\varpi\right) \\
&\leq \ldots \leq \eta^n\mathcal{G}(\varpi, f\varpi, f\varpi)
\end{aligned}$$

As $n>1$, so $\mathcal{G}(\varpi, f\varpi, f\varpi) \leq \eta\mathcal{G}(\varpi, f\varpi, f\varpi)$; that is, $\mathcal{G}(\varpi, f\varpi, f\varpi) = 0$, and $\varpi = f\varpi$. Thus $F\left(f^n\right) = F(f)$.

6.7 CONCLUSION

We studied fixed-point results in G-metric space, which is a general metric space. First two sections of the chapter deal with various topological properties and results dealing with the convergence and continuity of self-mappings in G-metric space. The third section deals with several fixed-point problems of self-maps in the setup of G-metric space. In the fourth section, we study various fixed-point theorems of mappings satisfying generalized weak contractions under the ordered G-metric structure. In the fifth section, we describe general T-Hardy Rogers–type contractions and study several fixed-point results for these maps under ordered G-metric spaces. In the sixth section, various periodic point results of generalized contractive maps in G-metric space are established. These results unify and extend many results of the literature.

6.8 FUTURE PERSPECTIVE AND SCOPE

G-metric space is a generalized metric space, and there are various important problems that need to be addressed in the future. Some of the suggested problems are the following:

(1) Study the best proximity point problem in the setup of G-metric space.
(2) Study coupled fixed-point results in the setup of G-metric space.
(3) Study fixed points of generalized contraction in G-metric space with a graph structure.

(4) Study the well-posedness of fixed points of generalized contraction in G -metric space.
(5) The stability of fixed-point problems on G -metric space.
(6) Fractals through iterated functions and multifunctions in a framework of G -metric space.

REFERENCES

[1] Dhage, B.C.: Generalized metric space and mapping with fixed point, *Bull. Cal. Math. Soc.* 84 (1992), 329–336.
[2] Dhage, B.C.: Generalized metric spaces and topological structure, I, *An. Stiint. Univ. Al.I. Cuza Iasi. Mat (N.S)* 46 (2000), 3–24.
[3] Dhage, B.C.: On generalized metric spaces and topological structure II, *Pure Appl. Math. Sci.* 40 (1–2) (1994), 37–41.
[4] Dhage, B.C.: On continuity of mappings in D-metric spaces, *Bull. Calcutta Math. Soc.* 86 (1994), 503–508.
[5] Mustafa, Z. and Sims, B.: Some remarks concerning D-metric spaces, in: Proc. Int. Conf. on Fixed Point Theor. Appl., Valencia, Spain, July 2003, pp. 189–198.
[6] Mustafa, Z. and Sims, B.: A new approach to generalized metric spaces, *J. Nonlinear Convex Anal.* 7 (2) (2006), 289–297.
[7] Mustafa, Z., Obiedat, H. and Awawdeh, F.: Some fixed point theorem for mapping on complete G–metric spaces, *Fixed Point Theory Appl.* 2008 (Article ID 189870) (2008), 12 pages.
[8] Mustafa, Z., Shatanawi, W. and Bataineh, M.: Existence of fixed point results in G–metric spaces, *Int. J. Math. Math. Sci.* 2009 (Article ID 283028) (2009), 10 pages.
[9] Mustafa, Z. and Sims, B.: Fixed point theorems for contractive mappings in complete G–metric spaces, *Fixed Point Theory Appl.* 2009 (Article ID 917175) (2009), 10 pages.
[10] Mustafa, Z., Awawdeh, F. and Shatanawi, W.: Fixed point theorem for expansive mappings in G– metric spaces, *Int. J. Contemp. Math. Sci.* 5 (2010), 2463–2472.
[11] Mustafa, Z., Khandagji, M. and Shatanawi, W.: Fixed point results on complete G–metric spaces, *Stud. Sci. Math. Hung.* 48 (3) (2011), 304–319.
[12] Saadati, R., Vaezpour, S. M., Vetro, P. and Rhoades, B.E.: Fixed point theorems in generalized partially ordered G– metric spaces, *Math. Comput. Model.* 52 (2010), 797–801.
[13] Abbas, M. and Nazir, T.: Fixed points of T–Hardy Rogers type contraction mapping in G–metric spaces, *Afr. Mat.* 25 (1) (2014), 103–113.
[14] Abbas, M., Nazir, T., Shatanawi, W. and Mustafa, Z.: Fixed and related fixed point theorems for three maps in G–metric spaces, *Hacet. J. Math. Stat.* 41 (2) (2012), 291–306.
[15] Abbas, M., Sintunavarat, W. and Kumam, P.: Coupled fixed point of generalized contractive mappings on partially ordered G–metric spaces, *Fixed Point Theory Appl.* 2012 (31), (2012).
[16] Gupta, V., Manro, S. and Verma, M.: A fixed point theorem in generalized metric spaces, *Sohag J. Math.* 2 (1) (2015), 37–39.
[17] Guran, L. and Latif, A.: Fixed point theorems for multivalued contractive operators on generalized metric spaces, *Fixed Point Theory* 16 (2) (2015), 327–336.
[18] Khan, S.H., Abbas, M. and Nazir, T.: Fixed point results for generalized Chatterjea type contractive conditions in partially ordered G–metric spaces, *Sci. World J.* 2014 (Article ID 341751) (2014), 8 pages.
[19] Shatanawi, W.: Fixed point theory for contractive mappings satisfying Φ –maps in G–metric spaces, *Fixed Point Theory Appl.* 2010 (Article ID 181650) (2010), 9 pages.

[20] Shatanawi, W., Abbas, M. and Nazir, T.: Common coupled fixed points results in two generalized metric spaces, *Fixed Point Theory Appl.* 2011 (80) (2011), 13 pages.

[21] Morales, J.R. and Rojas, E.: Cone metric spaces and fixed point theorems of T-Kannan contractive mappings, *Int. J. Math. Anal.* 4 (4) (2010), 175–184.

[22] Hicks, T.L. and Rhoades, B.E.: A Banach type fixed point theorem, *Math. Japonica* 24 (3) (1979), 327–330.

[23] Ciric, L. B.: On contraction type mappings, *Math. Balkanica* 1 (1971), 52–57.

[24] Park, S.: A unified approach to fixed points of contractive maps, *J. Korean Math. Soc.* 16 (1980), 95–105.

[25] Ran, A.C.M. and Reurings, M.C.B.: A fixed point theorem in partially ordered sets and some application to matrix equations, *Proc. Amer. Math. Soc.* 132 (2004), 1435–1443.

[26] Nieto, J.J. and Lopez, R.R.: Contractive mapping theorems in partially ordered sets and applications to ordinary differential equations, *Order* 22 (2005), 223–239.

[27] Gornicki, J. and Rhoades, B.E.: A general fixed point theorem for involutions, *Indian J. Pure Appl. Math.* 27 (1996), 13–23.

[28] Jeong, G.S. and Rhoades, B.E.: Maps for which $F(T)=F(T^n)$, *Fixed Point Theory* 6 (2005), 87–131.

[29] Singh, K.L.: Sequences of iterates of generalized contractions, *Fund. Math.* 105 (1980), 115–126.

[30] Abbas, M., Nazir, T. and Radenovic, S.: Some periodic point results in generalized metric spaces, *Appl. Math. Comput.* 217 (8) (2010), 4094–4099.

[31] Chugh, R., Kadian, T.A., Rani, A. and Rhoades, B.E.: Property P in G–metric spaces, *Fixed Point Theory Appl.* 2010 (ID 401684) (2010), 12 pages.

[32] Subrahmanyam, P.V.: Remarks on some fixed point theorems related to Banach contraction principle, *J. Math. Phys. Sci.* 8 (1974), 445–457.

[33] Choudhury, B.S. and Maity, P.: Coupled fixed point results in generalized metric spaces, *Math. Compu. Model.* 54 (2011), 73–79.

7 Inertial Three-Step Forward-Backward Splitting Algorithms to Solve Inclusion Problems and Their Application to Image Restoration Problems

W. Cholamjiak, D. Yambangwai, and S. Das

7.1 INTRODUCTION AND PRELIMINARIES

Let C be a nonempty closed convex subset of a real Hilbert H. A mapping $T : C \to C$ is said to be *nonexpansive* if

$$\| TJ - T\ell \| \leq \| J - \ell \|$$

for all $J, \ell \in C$.

A mapping $F : C \to H$ is called $\alpha -$ inverse strongly monotone if $\exists\ \alpha > 0$ such that

$$\langle J - \ell, FJ - F\ell \rangle \geq \alpha \| FJ - F\ell \|^2, \forall J, \ell \in C$$

It is well known that if $T : C \to C$ is nonexpansive, then $F \doteq I - T$ is $\frac{1}{2}$ — inverse strongly monotone (refer to [1, 2]). An operator $G : H \to 2^H$ is said to be monotone if $\langle J_1 - J_2, \ell_1 - \ell_2 \rangle \geq 0$ wherever $\ell_1 \in GJ_1$ and $\ell_2 \in GJ_2$. If the graph of a monotone operator G is not contained in the graph of any other monotone operator, it is said to be maximal.

In this chapter, we consider the problem of finding a solution to the inclusion problem by iteratively finding a zero of the sum of two monotone operators F and G in a Hilbert space H, where the inclusion problem is finding $\hat{J} \in H$ and

$$0 \in F\hat{J} + G\hat{J} \tag{7.1}$$

DOI: 10.1201/9781003322856-7

where $F: H \to H$ is an operator and $G: H \to 2^H$ is a multivalued operator. The inclusion problem (1.1) is well known to be equivalent to the fixed point equation

$$\hat{J} = (I + \tau G)^{-1}(\hat{J} - \tau F\hat{J})$$

Problem (7.1) encompasses a wide range of optimization problems, including convex programming, variational inequality, split feasibility, and minimization problems. It can be used to solve problems in machine learning, image processing, and the linear inverse problem, to be more explicit. The original forward-backward splitting algorithm (OFBS) [3–9] was one of the significant algorithms used to solve the problem (7.1). For $J_1 \in H$, the algorithm is defined as follows:

$$J_{n+1} = (I + \tau G)^{-1}(J_n - \tau F J_n), n \geq 1 \tag{7.2}$$

where $\tau > 0$. Every iteration step in this case basically involves F as the forward step and G as the backward step, not the sum of operators. Later on, Lions and Mercier [10] utilized a forward-backward splitting algorithm; it can be rewritten in a real Hilbert space as follows:

$$J_{n+1} = \mathfrak{J}_\tau^F \left(2\mathfrak{J}_\tau^G - I\right) J_n + \left(I - \mathfrak{J}_\tau^G\right) J_n, \; n \geq 1 \tag{7.3}$$

where $\mathfrak{J}_\tau^T = (I + \tau T)^{-1}$ with $\tau > 0$. This algorithm is called the Douglas-Rachford (DR) algorithm [11]. The weak convergence was proved by [12, 10].

The forward-backward splitting algorithm (FBS) was also presented in the book by Bauschke and Combettes [12]. The better thing than the Douglas-Rachford algorithm is easier to use in an application as it involves only a single operator G in J^G_\ tau, unlike Douglas-Rachford which requires two operators. The theorem is stated as follows:

$$\begin{cases} \ell_n = J_n - \tau F J_n \\ J_{n+1} = J_n + \alpha_n \left(\mathfrak{J}_\tau^G \ell_n - J_n\right) \end{cases} \tag{7.4}$$

Kitkuan et al. [13] introduced a novel viscosity (NV) iterative approach to obtain a strongly convergent theorem and applied it to image restoration. They were interested in a common zero of sum two operators, that is, $F^{-1}0 \cap G^{-1}0$ which different zero of sum of two operator, $(F+G)^{-1}(0)$. The algorithm is generated by an element J_0 in a nonempty closed and convex subset C of H, and

$$\begin{cases} \ell_n = \beta_n J_n + (1 - \beta_n) \mathfrak{J}_{\lambda_n}^G J_n \\ J_{n+1} = \alpha_n f(J_n) + (1 - \alpha_n) \mathfrak{J}_{\tau_n}^G \ell_n \end{cases} \tag{7.5}$$

where $f : C \to C$ is a contraction mapping. Defining the algorithm to obtain strong convergence is very important in the infinitely dimensional space. Two different hybrid projection algorithms were proposed by Takahashi et al. [14] and Nakajo and Takahashi [15].

One method that many researchers are interested in is speeding up the convergence of the algorithm. In 2011, Alvarez and Attouch citeAlvarez used the inertial technique for convergence, which was previously examined in [16, 17] for maximal monotone operators utilizing the proximal point algorithm by Alvarez and Attouch [18].1 After that, many mathematicians modified the inertial technique to better perform the algorithm and a nice convergent speed [19–22]. Recently, the proximal point algorithm, which is generated by the inertial technique and the forward-backward algorithm, was proposed by Moudafi and Oliny [23] to solve the zero-finding problem of the sum of two monotone operators. This algorithm was generated as follows:

$$\begin{cases} \ell_n = J_n + \theta_n (J_n - J_{n-1}) \\ J_{n+1} = (I + \tau_n G)^{-1} (\ell_n - \tau_n F J_n), n \geq 1 \end{cases} \tag{7.6}$$

The weak convergence theorem was obtained by the conditions that $\tau_n < 2/L$ where L is the Lipschitz constant of F and the control condition on the inertial term. After that, many mathematicians used the inertial technique for modifying the algorithm [7].

We introduce an inertial three-step forward-backward splitting method for approximating an element of the set of solutions to the inclusion problem (7.1) and prove the weak convergence theorem under acceptable conditions, which is motivated and inspired by the existing literature. Moreover, we present two different algorithms by using the hybrid projection algorithms to obtain strongly convergent theorems. We then give an example of the main theorems and compare our proposed inertial projection method and the existing standard projection method. In the final section, we implement our methods to solve unconstrained image recovery problems and compare them to the OFBS (1.2), DR(1.3), FBS(1.6), and NV (1.9) algorithms. It's compelling that our method has a significantly better rate of convergence.

7.2 MAIN RESULTS

In this section, we prove weak and strong convergence. For this purpose, we use a three-step forward-backward splitting method for solving inclusion problems and prove the weak convergence theorem for a variation of the inertial technical term. Moreover, using two different hybrid projection methods, we prove the strong convergence results.

Theorem 7.1. *Let H be a real Hilbert space, an α-inverse strongly monotone operator $F : H \to H$, and a maximal monotone operator $G : H \to 2^H$ such that $(F+G)^{-1}(0) \neq \varnothing$. Let $\{J_n\}$ be a sequence defined by $J_0, J_1 \in H$ and*

$$\begin{cases} \ell_n &= J_n + \theta_n (J_n - J_{n-1}) \\ r_n &= \gamma_n \ell_n + (1-\gamma_n) \Im_{\tau_n}^G (I - \tau_n F) \ell_n \\ s_n &= \beta_n r_n + (1-\beta_n) \Im_{\tau_n}^G (I - \tau_n F) r_n \\ J_{n+1} &= \alpha_n s_n + (1-\alpha_n) \Im_{\tau_n}^G (I - \tau_n F) s_n \end{cases} \tag{7.7}$$

where $\Im_{\tau_n}^G = (I + \tau_n G)^{-1}$, $\{\tau_n\} \subset (0, 2\alpha)$, $\{\theta_n\} \subset [0, \theta]$ *for some* $\theta \in [0,1)$ *and the real sequences* $\{\alpha_n\}$, $\{\beta_n\}$, $\{\gamma_n\} \in [0,1]$. *Assume that*

(i) $\sum_{n=1}^{\infty} \theta_n \| J_n - J_{n-1} \| < \infty$;

(ii) $\limsup J_{n \to \infty} \gamma_n < 1$;

(iii) $0 < \liminf_{n \to \infty} \tau_n \leq \limsup J_{n \to \infty} \tau_n < 2\alpha$.

Then the sequence $\{J_n\} \rightharpoonup \hat{J}$ *where* $\hat{J} \in (F+G)^{-1}(0)$.

Proof. Let $\hat{J} \in (F+G)^{-1}(0)$ and $\forall \ n \in \mathbb{N}$, $\Im_n = (I + \tau_n G)^{-1}(I - \tau_n F)$. Since $\Im_n$ is nonexpansive, we have

$$\begin{aligned} \| J_{n+1} - \hat{J} \| &\leq \alpha_n \| s_n - \hat{J} \| + (1-\alpha_n) \| \Im_n s_n - \hat{J} \| \\ &\leq \| s_n - \hat{J} \| \\ &\leq \beta_n \| r_n - \hat{J} \| + (1-\beta_n) \| \Im_n r_n - \hat{J} \| \\ &\leq \| r_n - \hat{J} \| \\ &\leq \gamma_n \| \ell_n - \hat{J} \| + (1-\gamma_n) \| \Im_n \ell_n - \hat{J} \| \\ &\leq \| \ell_n - \hat{J} \| \\ &\leq \| J_n - \hat{J} \| + \theta_n \| J_n - J_{n-1} \| \end{aligned} \tag{7.8}$$

It follows from the Theorem 7.1 of [18] and assumption (i) that $\lim_{n \to \infty} \| J_n - \hat{J} \|$ exists and the sequence $\{J_n\}$ is bounded. It now follows from the boundedness of $\{J_n\}$ that $\{\ell_n\}$, $\{r_n\}$, and $\{s_n\}$ are bounded. By Lemma 7.3 of [6], we conclude that

$$\begin{aligned} \| J_{n+1} - \hat{J} \|^2 &\leq \alpha_n \| s_n - \hat{J} \|^2 + (1-\alpha_n) \| \Im_n s_n - \hat{J} \|^2 \\ &\leq \| s_n - \hat{J} \|^2 \\ &\leq \beta_n \| r_n - \hat{J} \|^2 + (1-\beta_n) \| \Im_n r_n - \hat{J} \|^2 \\ &\leq \| r_n - \hat{J} \|^2 \\ &\leq \gamma_n \| \ell_n - \hat{J} \|^2 + (1-\gamma_n) \| \Im_n \ell_n - \hat{J} \|^2 \end{aligned}$$

$$\leq \| \ell_n - \hat{J} \|^2 - (1-\gamma_n)$$
$$\left(\tau_n (2\alpha - \tau_n) \| F\ell_n - F\hat{J} \|^2 + \| \ell_n - \tau_n F\ell_n - \Im_n \ell_n + \tau_n F\hat{J} \|^2\right).$$
$$\leq \| J_n - \hat{J} \|^2 + 2\theta_n \left\langle J_n - J_{n-1}, \ell_n - \hat{J} \right\rangle - (1-\gamma_n)\left(\tau_n (2\alpha - \tau_n) \| F\ell_n - F\hat{J} \|^2\right.$$
$$\left. + \| \ell_n - \tau_n F\ell_n - \Im_n \ell_n + \tau_n F\hat{J} \|^2\right) \tag{7.9}$$

This implies that

$$(1-\gamma_n)\left(\tau_n (2\alpha - \tau_n) \| F\ell_n - F\hat{J} \|^2 + \| \ell_n - \tau_n F\ell_n - \Im_n \ell_n + \tau_n F\hat{J} \|^2\right)$$
$$\leq \| J_n - \hat{J} \|^2 - \| J_{n+1} - \hat{J} \|^2 + 2\theta_n \left\langle J_n - J_{n-1}, \ell_n - \hat{J} \right\rangle \tag{7.10}$$

It now follows from existence of $\lim_{n\to\infty} \| J_n - p \|$, (2), and assumptions (i)–(iii) that

$$\lim_{n\to\infty} \| F\ell_n - F\hat{J} \| = \lim_{n\to\infty} \| \ell_n - \tau_n F\ell_n - \Im_n \ell_n + \tau_n F\hat{J} \| = 0 \tag{7.11}$$

We can deduce from the triangle inequality that

$$\lim_{n\to\infty} \| \Im_n \ell_n - \ell_n \| = 0 \tag{7.12}$$

Since $\liminf_{n\to\infty} \tau_n > 0 \ \exists \ \tau > 0$ such that $\tau_n \geq \tau \ \ \forall \ \ n \geq 1$. Lemma 7.2 of [6] implies that

$$\| T_\tau^{F,G} \ell_n - \ell_n \| \leq 2 \| \Im_n \ell_n - \ell_n \| \tag{7.13}$$

Then, by (7.6) and (7.7), we obtain

$$\lim_{n\to\infty} \| T_\tau^{F,G} \ell_n - \ell_n \| = 0 \tag{7.14}$$

It now follows from the definition of $\{J_n\}$ and assumption (i) that

$$\lim_{n\to\infty} \| \ell_n - J_n \| = \lim_{n\to\infty} \theta_n \| J_n - J_{n-1} \| = 0 \tag{7.15}$$

Notice that H is reflexive and $\{J_n\}$ is bounded; $\omega_w(J_n) = \left\{\hat{J} \in H : J_{n_i} \rightharpoonup p, \{J_{n_i}\} \subset \{J_n\}\right\}$ is nonempty. For an arbitrary element $q \in \omega_w(J_n)$ $\exists$, a subsequence $\{J_{n_i}\} \subset \{J_n\}$ $\rightharpoonup q$. Let $\hat{J} \in \omega_w(J_n)$ and $\{J_{n_m}\} \subset \{J_n\} \ni J_{n_m} \rightharpoonup \hat{J}$. Then from (7.9), $\ell_{n_i} \rightharpoonup q$ and $\ell_{n_m} \rightharpoonup \hat{J}$. In view of the fact that $T_\tau^{F,G} \rightharpoonup$ is nonexpansive, it results from the demiclosedness of a nonexpansive mapping [24] that $\hat{J}, q \in (F+G)^{-1}(0)$. Also, due to the fact that real Hilbert space satisfies Opial's condition, we are allowed to use Lemma 7.7 in [25] to obtain $\hat{J} = q$.

Theorem 7.2. *Let a real Hilbert space H, an α-inverse strongly monotone operator $F: H \to H$, and a maximal monotone operator $G: H \to 2^H \ni (F+G)^{-1}(0) \neq \varnothing$. Let a sequence $\{J_n\}$ be defined by $C_1 = H$, $J_0, J_1 \in H$ and*

$$\begin{cases} \ell_n &= J_n + \theta_n(J_n - J_{n-1}) \\ r_n &= \gamma_n \ell_n + (1-\gamma_n)\mathfrak{I}^G_{\tau_n}(I - \tau_n f)\ell_n \\ s_n &= \beta_n r_n + (1-\beta_n)\mathfrak{I}^G_{\tau_n}(I - \tau_n f) r_n \\ t_n &= \alpha_n s_n + (1-\alpha_n)\mathfrak{I}^G_{\tau_n}(I - \tau_n f) s_n \\ C_{n=1} &= \{z \in C_n : \| t_n - z \|^2 \leq \| J_n - z \|^2 + 2\theta_n^2 \| J_n - J_{n-1} \|^2 - 2\theta_n \langle J_n - z, J_{n-1} - J_n \rangle\} \\ J_{n+1} &= P_{C_{n+1}} J_n, n \geq 1 \end{cases} \tag{7.16}$$

where $\mathfrak{I}^G_{\tau_n} = (I + \tau_n G)^{-1}$, $\{\tau_n\} \subset (0, 2\alpha)$, $\{\theta_n\} \subset [0, \theta]$ where $\theta \in [0,1)$, and the real sequences $\{\alpha_n\}$, $\{\beta_n\}$, $\{\gamma_n\} \in [0,1]$. Assume that

(i) $\sum_{n=1}^{\infty} \theta_n \| J_n - J_{n-1} \| < \infty$;

(ii) $\limsup J_{n\to\infty} \gamma_n < 1$;

(iii) $0 < \liminf_{n\to\infty} \tau_n \leq \limsup J_{n\to\infty} \tau_n < 2\alpha$.

Then the sequence $\{J_n\}$ converges strongly to $\hat{J} = P_{(F+G)^{-1}(0)} J_1$.

Proof. For proof, we proceed in five steps, result of which will be used in the sequel.

Step 1: To prove that $\forall\ \hat{J} \in H$, $P_{C_{n+1}} J_1$ is well defined. In view of Lemma 7.1 of [6], $(F+G)^{-1}(0)$ is closed and convex.

It now follows from Lemma 7.3 of [26] and the definition of C_{n+1} that $\forall\ n \geq 1$, and C_{n+1} is closed and convex where $n \in \mathbb{N}$. Let $\mathfrak{I}_n = (I + \tau_n G)^{-1}(I - \tau_n F)$ and let $\hat{J} \in (F+G)^{-1}(0)$. Notice that $\mathfrak{I}_n$ is nonexpansive; we have

$$\begin{aligned} \| t_n - \hat{J} \|^2 &\leq \alpha_n \| s_n - \hat{J} \|^2 + (1-\alpha_n) \| \mathfrak{I}_n s_n - \hat{J} \|^2 \\ &\leq \| s_n - \hat{J} \|^2 \\ &\leq \beta_n \| r_n - \hat{J} \|^2 + (1-\beta_n) \| \mathfrak{I}_n r_n - \hat{J} \|^2 \\ &\leq \| r_n - \hat{J} \|^2 \\ &\leq \gamma_n \| \ell_n - \hat{J} \|^2 + (1-\gamma_n) \| \mathfrak{I}_n \ell_n - \hat{J} \|^2 \\ &\leq \| \ell_n - \hat{J} \|^2 \\ &\leq \| J_n - \hat{J} \|^2 + 2\theta_n \langle J_n - J_{n-1}, \ell_n - \hat{J} \rangle \\ &\leq \| J_n - \hat{J} \|^2 + 2\theta_n^2 \| J_n - J_{n-1} \|^2 - 2\theta_n \langle J_n - \hat{J}, J_{n-1} - J_n \rangle \end{aligned} \tag{7.17}$$

So, we have $p \in C_{n+1}$, thus $(F+G)^{-1}(0) \subset C_{n+1}$. Therefore $P_{C_{n+1}} J_1$ is well defined.

Step 2: To prove $\lim_{n\to\infty} \| J_n - J_1 \|$ exists, consider the fact that nonempty $(F+G)^{-1}(0) \subseteq H$ is closed and convex; it follows that $\exists$ a unique $v \in S \ni$

$$v = P_{(F+G)^{-1}(0)} J_1 \tag{7.18}$$

We can deduce from $J_n = P_{C_n} J_1$, $C_{n+1} \subset C_n$ and $J_{n+1} \in C_{n+1}$, $\forall n \geq 1$ that

$$\| J_n - J_1 \| \leq \| J_{n+1} - J_1 \|, \forall n \geq 1 \tag{7.19}$$

However, the fact that $(F+G)^{-1}(0) \subset C_n$ results in

$$\| J_n - J_1 \| \leq \| v - J_1 \|, \forall n \geq 1 \tag{7.20}$$

It implies that the sequence $\{J_n\}$ is bounded and nondecreasing, so it now follows that $\lim_{n\to\infty} \| J_n - J_1 \|$ exists.

Step 3: To prove that $J_n \to p \in C$ as $n \to \infty$, we follow the definition of C_n for $m > n$, which yields $J_m = P_{C_m} J_1 \in C_m \subseteq C_n$. In the sense of metric projection properties of the P_{C_n}, we have

$$\| J_m - J_n \|^2 \leq \| J_m - J_1 \|^2 - \| J_n - J_1 \|^2 \tag{7.21}$$

Since $\lim_{n\to\infty} \| J_n - J_1 \|$ exists, it follows from (7.21) that $\lim_{n\to\infty} \| J_m - J_n \| = 0$. Hence $\{J_n\} \in C$ is a Cauchy sequence, and hence for $q \in C$, $J_n \to q$ as $n \to \infty$.

Step 4: To prove that $p \in (F+G)^{-1}(0)$. Equivalently, by Step 3, we obtain $\lim_{n\to\infty} \| J_{n+1} - J_n \| = 0$. From the fact $J_{n+1} \in C_{n+1}$, it results that

$$\begin{aligned} \| t_n - J_n \| &\leq \| t_n - J_{n+1} \| + \| J_{n+1} - J_n \| \\ &\leq \sqrt{\| J_n - J_{n+1} \|^2 + 2\theta_n^2 \| J_n - J_{n-1} \|^2 - 2\theta_n \langle J_n - J_{n+1}, J_{n-1} - J_n \rangle} \\ &\quad + \| J_{n+1} - J_n \| \end{aligned} \tag{7.22}$$

By assumption (i) and (7.22), we have

$$\lim_{n\to\infty} \| t_n - J_n \| = 0 \tag{7.23}$$

By Lemma 7.3 of [6] and (7.17), we deduce that

$$\begin{aligned} &\| t_n - \hat{J} \|^2 \leq \gamma_n \| \ell_n - \hat{J} \|^2 + (1-\gamma_n) \| \Im_n \ell_n - \hat{J} \|^2 \\ &\leq \| \ell_n - \hat{J} \|^2 - (1-\gamma_n)\left(\tau_n (2\alpha - \tau_n) \| F\ell_n - F\hat{J} \|^2 + \| \ell_n - \tau_n F\ell_n - \Im_n \ell_n + \tau_n F\hat{J} \|^2\right) \\ &\leq \| J_n - \hat{J} \|^2 + 2\theta_n \langle \hat{J}_n - J_{n-1}, \ell_n - \hat{J} \rangle - (1-\gamma_n)\left(\tau_n (2\alpha - \tau_n) \| F\ell_n - F\hat{J} \|^2 \right. \\ &\qquad \left. + \| \ell_n - \tau_n F\ell_n - \Im_n \ell_n + \tau_n F\hat{J} \|^2\right) \end{aligned} \tag{7.24}$$

This implies that

$$(1-\gamma_n)\left(\tau_n(2\alpha-\tau_n)\|F\ell_n - F\hat{J}\|^2 + \|\ell_n - \tau_n F\ell_n - \Im_n\ell_n + \tau_n F\hat{J}\|^2\right)$$
$$\leq \|J_n - \hat{J}\|^2 - \|t_n - \hat{J}\|^2 + 2\theta_n\left\langle J_n - J_{n-1}, \ell_n - \hat{J}\right\rangle \tag{7.25}$$

From (7.23); (2); and assumptions (i), (ii), and (iii), we have

$$\lim_{n\to\infty}\|F\ell_n - F\hat{J}\| = \lim_{n\to\infty}\|\ell_n - \tau_n F\ell_n - \Im_n\ell_n + \tau_n F\hat{J}\| = 0 \tag{7.26}$$

which deduces from the triangle inequality that

$$\lim_{n\to\infty}\|\Im_n\ell_n - \ell_n\| = 0 \tag{7.27}$$

It is given that $\liminf_{n\to\infty}\tau_n > 0 \;\exists\; \tau > 0$ such that $\tau_n \geq \tau \;\;\forall\;\; n \geq 1$. By using Lemma 7.2 of [6], we have

$$\|T_\tau^{F,G}\ell_n - \ell_n\| \leq 2\|\Im_n\ell_n - \ell_n\| \tag{7.28}$$

Then, by (7.27) and (7.28), we obtain

$$\lim_{n\to\infty}\|T_\tau^{F,G}\ell_n - \ell_n\| = 0 \tag{7.29}$$

From (7.28), we have

$$\lim_{n\to\infty}\|r_n - \ell_n\| = \lim_{n\to\infty}(1-\gamma_n)\|\Im_n\ell_n - \ell_n\| = 0 \tag{7.30}$$

By assumption (i), we have

$$\lim_{n\to\infty}\|\ell_n - J_n\| = \lim_{n\to\infty}\theta_n\|J_n - J_{n-1}\| = 0 \tag{7.31}$$

From Step 3, we know that $J_n \to \hat{J} \in C$. It follows from (7.31) that $\ell_n \to p$. This implies that $\hat{J} \in (F+G)^{-1}(0)$ by the demiclosedness of nonexpansive mapping $T_\tau^{F,G}$ and (7.29).

Step 5. To prove that $\hat{J} = P_{(F+G)^{-1}(0)}J_1$, we are given that $J_n = P_{C_n}J_1$ and $(F+G)^{-1}(0) \subset C_n$. It results

$$\langle J_1 - J_n, J_n - z\rangle \geq 0, \forall z \in (F+G)^{-1}(0) \tag{7.32}$$

which gives

$$\left\langle J_1 - \hat{J}, \hat{J} - z\right\rangle \geq 0, \forall z \in (F+G)^{-1}(0) \tag{7.33}$$

whenever $n \to \infty$ in (7.32) and hence $\hat{J} = P_{(F+G)^{-1}(0)}J_1$.

Theorem 7.3. *Let a real Hilbert space* H*, an* α *-inverse strongly monotone operator* $f: H \to H$*, and a maximal monotone operator* $G: H \to 2^H \ni (f+G)^{-1}(0) \neq \phi$. *Let a sequence* $\{J_n\}$ *be defined by* $C_1 = Q_1 = H$, $J_0, J_1 \in H$ *and*

$$\begin{cases} \ell_n & = J_n + \theta_n (J_n - J_{n-1}) \\ r_n & = \gamma_n \ell_n + (1-\gamma_n) \Im_{\tau_n}^G (I - \tau_n F) \ell_n \\ s_n & = \beta_n r_n + (1-\beta_n) \Im_{\tau_n}^G (I - \tau_n F) r_n \\ t_n & = \alpha_n s_n + (1-\alpha_n) \Im_{\tau_n}^G (I - \tau_n F) s_n \\ C_n & = \{ z \in H : \| t_n - z \|^2 \leq \| J_n - z \|^2 + 2\theta_n^2 \| J_n - J_{n-1} \|^2 - 2\theta_n \langle J_n - z, J_{n-1} - J_n \rangle \} \\ \ell_n & = \{ z \in \ell_{n-1} : \langle J_1 - J_n, J_n - z \rangle \geq 0 \} \\ J_{n+1} & = P_{C_n \cap \ell_n} J_1, n \geq 1 \end{cases} \tag{7.34}$$

where $\Im_{\tau_n}^G = (I + \tau_n G)^{-1}$, $\{\tau_n\} \subset (0, 2\alpha)$, $\{\theta_n\} \subset [0, \theta]$ where $\theta \in [0,1)$ and the real sequences $\{\alpha_n\}$, $\{\beta_n\}$, $\{\gamma_n\} \in [0,1]$. Assume that

(i) $\sum_{n=1}^{\infty} \theta_n \| J_n - J_{n-1} \| < \infty$;

(ii) $\limsup J_{n \to \infty} \gamma_n < 1$;

(iii) $0 < \liminf_{n \to \infty} \tau_n \leq \limsup J_{n \to \infty} \tau_n < 2\alpha$.

Then the sequence $\{J_n\}$ *converges strongly to* $\hat{J} = P_{(F+G)^{-1}(0)} J_1$.

Proof. In view of the proof of Theorem 7.2, we can deduce the proof by replacing C_{n+1} with C_n, except in Step 7.1. This shows that $(f+G)^{-1}(0) \subseteq C_n \;\; \forall \;\; n \geq 1$. Next we show that $(f+G)^{-1}(0) \subseteq \ell_n \;\; \forall \;\; n \in \mathbb{N}$. For $n = 1$, the induction results in $(f+G)^{-1}(0) \subseteq H = Q_1$. Let $(f+G)^{-1}(0) \subseteq \ell_n \;\; \forall \;\; n \in \mathbb{N}$. Recall that J_{n+1} is the projection of J_1 onto $C_n \cap \ell_n$; we have

$$\langle J_1 - J_{n+1}, J_{n+1} - y \rangle \geq 0J, \forall y \in C_n \cap \ell_n \tag{7.35}$$

Thus $(f+G)^{-1}(0) \subseteq \ell_{n+1}$, so $(f+G)^{-1}(0) \subseteq C_n \cap \ell_n$. This implies that $\{J_n\}$ is well defined. The definition of ℓ_n results $J_n = P_{\ell_n} J_1$. We have inequality (2.13) because $J_{n+1} \in \ell_n$. Contrarily, we have

$$\| J_n - J_1 \| \leq \| z - J_1 \|, \forall z \in (f+G)^{-1}(0) \tag{7.36}$$

Inequalities (7.19) and (7.36) imply that the sequence $\{J_n - J_1\}$ is bounded and nondecreasing; hence $\lim_{n\to\infty} \| J_n - J_1 \|$ exists. Also, the definition of ℓ_n results in $J_m = P_{Q_m} J_1 \in Q_m \subseteq \ell_n$ for $m > n$. In the sense of metric projection properties P_{Q_m}, we obtain the equation (7.21). Since $\lim_{n\to\infty} \| J_n - J_1 \|$ exists, it follows from (7.21) that $\lim_{n\to\infty} \| J_m - J_n \| = 0$. Hence $\{J_n\}$ is a Cauchy sequence, and $J_n \to \hat{J}$ as $n \to \infty$ for some $\hat{J} \in H$. In particular, we have $\lim_{n\to\infty} \| J_{n+1} - J_n \| = 0$. Since $J_{n+1} \in C_n$, we obtain inequality (7.22). It follows from the proof of Steps 4–5 in Theorem 7.2 that we obtain $\hat{J} = P_{(f+G)^{-1}(0)} J_1$.

Let us illustrate the fact for strongly convergent theorems in infinitely dimensional spaces with the following example which is functional space $L_2[0,1]$ to support the main theorems Theorem 7.2 and 7.3.

Example 7.4. For $H = L_2[0,1]$, let us define the Volterra integral operator $F : L_2[0,1] \to L_2[0,1]$ by $Fp(t) = \int_0^t p(s)ds,$ and let $G : L_2[0,1] \to L_2[0,1]$ be defined by $Gp(t) = 5p(t)$, where $p(t) \in L_2[0,1]$, $t \in [0,1]$. Then F is $\frac{\pi}{2}$-inverse strongly monotone, and G is maximal monotone. With direct calculation, we have for

$$\begin{aligned} \Im_{\tau_n}^{G}(p - \tau_n Fp) &= (I + \tau_n G)^{-1}(p - \tau_n Fp) \\ &= \frac{p(t) - \tau_n \int_0^t p(s)ds}{1 + 5\tau_n} \end{aligned} \tag{7.37}$$

where $\tau_n > 0$ and $p(t) \in L_2[0,1]$, $t \in [0,1]$. Since $\alpha = \frac{\pi}{2}$, we can choose $\tau_n = 1\,\forall\, n \in \mathbb{N}$. Let $\alpha_n = \beta_n = \gamma_n = 1 - \frac{1}{100n^2 + 1}$ and

$$\theta_n = \begin{cases} \min\left\{ \frac{1}{n^2 \| J_n - J_{n-1} \|}, 0.25 \right\} & if\ J_n \neq J_{n-1} \\ 0.25 & otherwise \end{cases}$$

It's easy to see that the sequence θ_n satisfies condition (i). A comparison test between our inertial forward-backward method defined in Theorem 7.2–7.3 and a standard forward-backward method (i.e. $\theta_n = 0$) are shown by the stopping criterion $\| J_n - J_{n-1} \| < 10^{-4}$.

The different choices of J_0 and J_1 are given in Table 7.1 as follows:

Choice 1 Bezier initial data: $J_0(t) = -10t^9(t-1)$ and $J_1(t) = -120t^7(t-1)^3$;

Choice 2 Chebyshev initial data: $J_0(t) = 9t - 120t^3 + 432t^5 - 576t^7 + 256t^9$ and $J_1(t) = t(56t^2 - 112t^4 + 64t^6 - 7)$;

Choice 3 Legendre initial data: $J_0(t) = 0.0078125t(18018t^4 - 4620t^2 - 25740t^6 + 12155t^8 + 315)$ and $J_1(t) = 0.0625t(315t^2 - 693t^4 + 429t^6 - 35)$.

Table 7.1
Comparison of $\theta_n \neq 0$ and $\theta_n = 0$ in Example 7.4

	Iteration (2.10)				Iteration (2.28)			
Choice of	**Iteration Number**		**CPU Time (sec)**		**Iteration Number**		**CPU Time (sec)**	
Initial Data	$\theta_n = 0$	$\theta_n \neq 0$	$\theta_n = 0$	$\theta_n \neq 0$	$\theta_n = 0$	$\theta_n \neq 0$	$\theta_n = 0$	$\theta_n \neq 0$
1	11	10	6.457101	5.762477	11	10	7.687459	7.282041
2	14	11	7.421536	6.327474	14	11	10.77674	8.908637
3	12	10	5.936844	5.885950	12	10	9.358310	8.040968

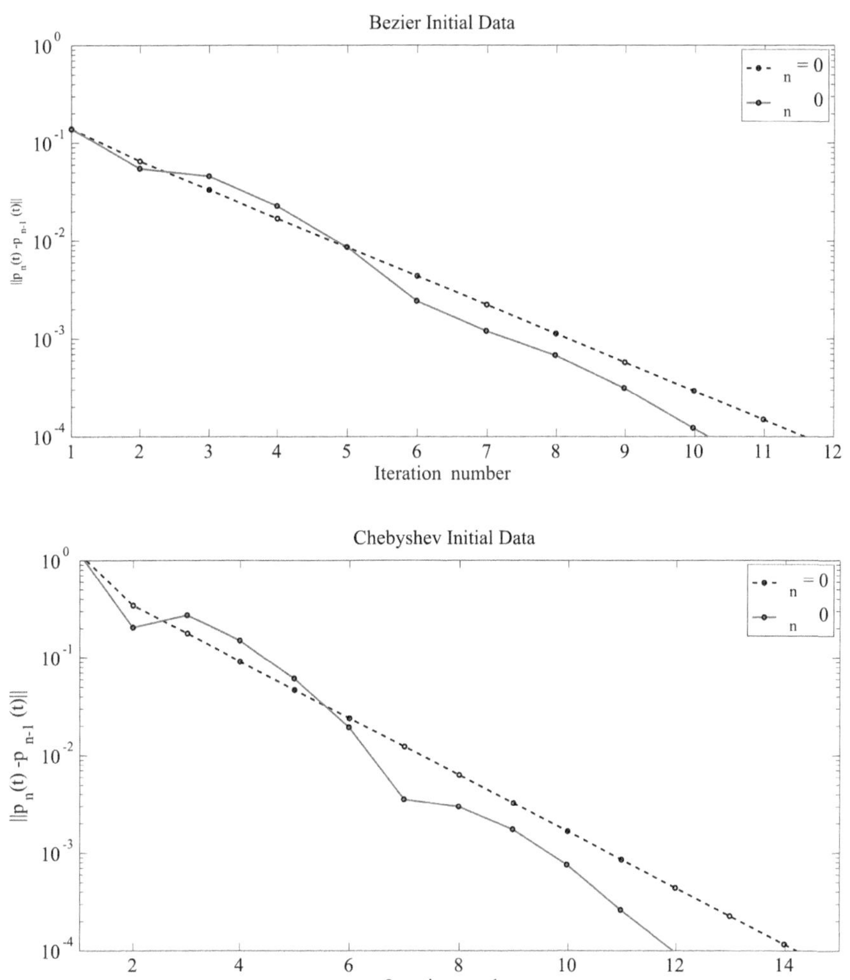

FIGURES 7.1–7.3 The error plotting $\| J_n - J_{n-1} \|$ with $\theta_n \neq 0$. and $\theta_n = 0$ for the inclusion problem with choices $1-3$ for initial data of Iteration (2.10) from Table 7.1.

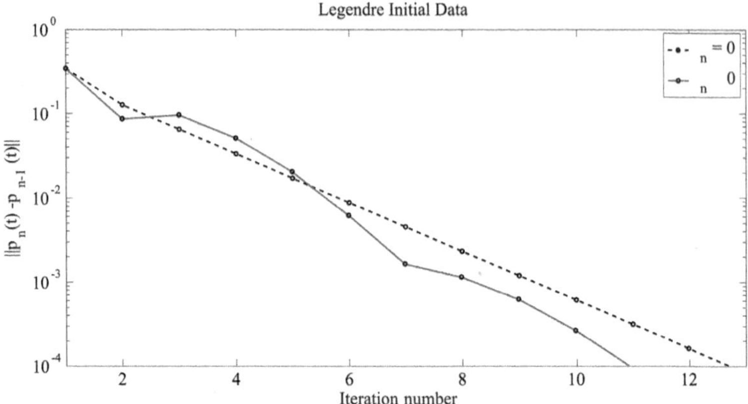

FIGURES 7.1–7.3 (Continued)

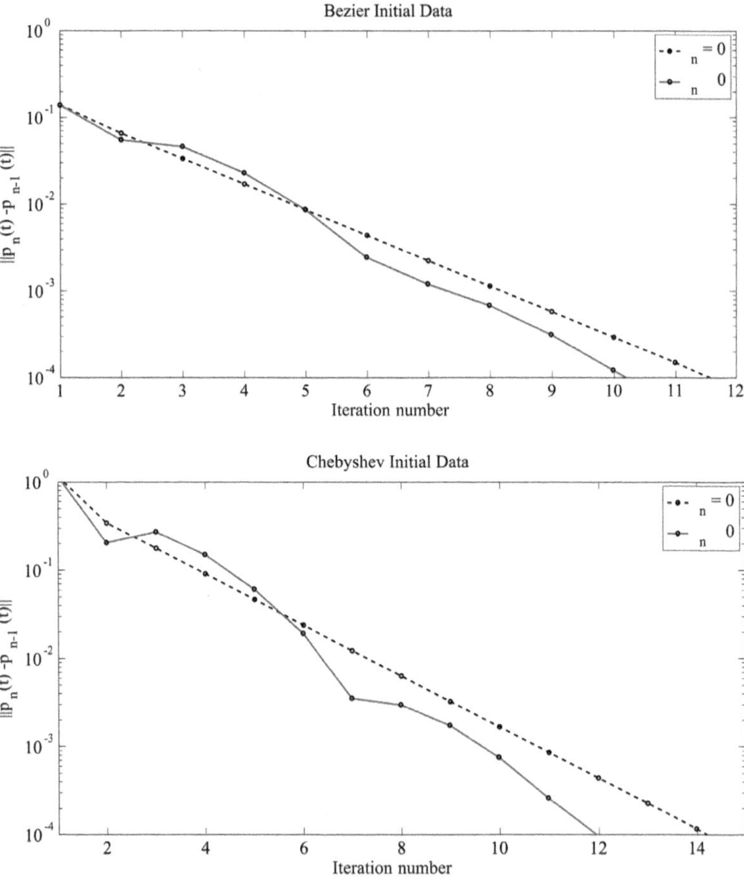

FIGURES 7.4–7.6 The error plotting $\| J_n - J_{n-1} \|$ with $\theta_n \neq 0$ and $\theta_n = 0$ for the inclusion problem with choices $1-3$ for initial data of Iteration (2.28) from Table 7.1.

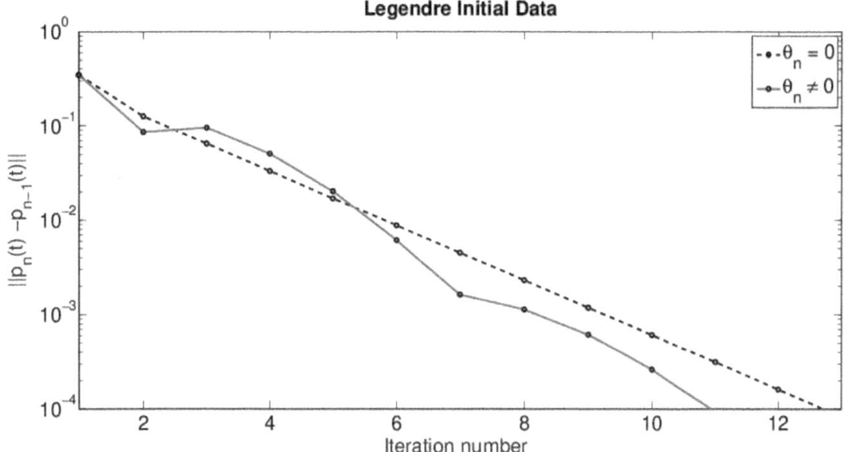

FIGURES 7.4–7.6 (Continued)

Remark 7.5. Figures 7.1–7.6 demonstrate that our three-step forward-backward hybrid projection methods [(7.16), (7.34)] are effective. For each of the options, using the inertial technical term results in faster convergence and fewer iterations than the standard forward-backward method ($\theta n = 0$).

7.3 SOME APPLICATIONS TO OPTIMIZATION PROBLEMS

In this section, we will discuss how our algorithms can be used to solve minimization problems and variational inequalities.

7.3.1 Variational Inequality Problem

For a nonlinear monotone operator $F: C \to H$, the variational inequality problem (VIP) is to find a point $\hat{p} \in C \ni$

$$\langle F\hat{p}, p - \hat{p} \rangle \geq 0, \forall p \in C \tag{7.38}$$

Let S represents the solution set of (7.38). It is important to no that solution of VIP (7.38) can be found with an extragradient method. It is well known that p solves VI(A,C) if and only if p solves the fixed-point equation (see [27] for details)

$$p = P_C(p - \lambda Fp), \lambda > O$$

Now the results are straightforward consequences of the previous application.

Theorem 7.1. *Let a real Hilbert space H, an α-inverse strongly monotone operator $F: H \to H$, a nonempty closed convex subset C of H, and $S \neq \varnothing$. Let a sequence $\{J_n\}$ be defined by $C_1 = H$, $J_0, J_1 \in H$ and*

$$\begin{cases} \ell_n & = J_n + \theta_n (J_n - J_{n-1}) \\ r_n & = \gamma_n \ell_n + (1-\gamma_n) P_C(\ell_n - \tau_n F\ell_n) \\ s_n & = \beta_n r_n + (1-\beta_n) P_C(r_n - \tau_n F r_n) \\ t_n & = \alpha_n s_n + (1-\alpha_n) P_C(s_n - \tau_n F s_n) \\ C_{n+1} & = \{z \in C_n : \| t_n - z \|^2 \leq \| J_n - z \|^2 + 2\theta_n^2 \| J_n - J_{n-1} \|^2 - 2\theta_n \langle J_n - z, J_{n-1} - J_n \rangle\} \\ J_{n+1} & = P_{C_{n+1}} J_1, n \geq 1 \end{cases} \tag{7.39}$$

where $\{\tau_n\} \subset (0, 2\alpha)$, $\{\theta_n\} \subset [0, \theta]$ for some $\theta \in [0, 1)$ and $\{\alpha_n\}$, $\{\beta_n\}$, $\{\gamma_n\}$ are sequences in $[0, 1]$. Assume that

(i) $\sum_{n=1}^{\infty} \theta_n \| J_n - J_{n-1} \| < \infty$;

(ii) $\limsup J_{n \to \infty} \gamma_n < 1$;

(iii) $0 < \liminf_{n \to \infty} \tau_n \leq \limsup J_{n \to \infty} \tau_n < 2\alpha$.

Then the sequence $\{J_n\}$ converges strongly to $\hat{J} = P_S J_1$.

Theorem 7.2. Let a real Hilbert space H, an α-inverse strongly monotone operator $F: H \to H$, a nonempty closed convex subset C of H, and $S \neq \varnothing$. Let a sequence $\{J_n\}$ be defined by $C_1 = Q_1 = H$, $J_0, J_1 \in H$ and

$$\begin{cases} \ell_n & = J_n + \theta_n (J_n - J_{n-1}) \\ r_n & = \gamma_n \ell_n + (1-\gamma_n) P_C(\ell_n - \tau_n F\ell_n) \\ s_n & = \beta_n r_n + (1-\beta_n) P_C(r_n - \tau_n F r_n) \\ t_n & = \alpha_n s_n + (1-\alpha_n) P_C(s_n - \tau_n F s_n) \\ C_n & = \{z \in H : \| t_n - z \|^2 \leq \|^2 J_n - z \| + 2\theta_n^2 \| J_n - J_{n-1} \|^2 - 2\theta_n \langle J_n - z, J_{n-1} - J_n \rangle\} \\ \ell_n & = \{z \in \ell_{n-1} : \langle J_n - J_n, J_n - z \rangle \geq 0\} \\ J_{n+1} & = P_{C_n \cap \ell_n} J_1, n \geq 1 \end{cases} \tag{7.40}$$

where $\{\tau_n\} \subset (0, 2\alpha)$, $\{\theta_n\} \subset [0, \theta]$ for some $\theta \in [0, 1)$ and $\{\alpha_n\}$, $\{\beta_n\}$, $\{\gamma_n\}$ are sequences in $[0, 1]$. Assume that

(i) $\sum_{n=1}^{\infty} \theta_n \| J_n - J_{n-1} \| < \infty$;

(ii) $\limsup J_{n \to \infty} \gamma_n < 1$;

(iii) $0 < \liminf_{n \to \infty} \tau_n \leq \limsup J_{n \to \infty} \tau_n < 2\alpha$.

Then the sequence $\{J_n\}$ converges strongly to $\hat{J} = P_S J_1$.

Table 7.2
Comparison of $\theta_n \neq 0$ and $\theta_n = 0$ in Example 7.3

	Iteration (3.2)				Iteration (3.3)			
Choice of	**Iteration Number**		**CPU Time (sec)**		**Iteration Number**		**CPU Time (sec)**	
Initial Data	$\theta_n = 0$	$\theta_n \neq 0$	$\theta_n = 0$	$\theta_n \neq 0$	$\theta_n = 0$	$\theta_n \neq 0$	$\theta_n = 0$	$\theta_n \neq 0$
1	100	92	326.7676	256.4288	100	92	761.9792	700.3430
2	100	100	595.6064	493.1987	100	100	524.2744	360.4290
3	100	100	784.0669	666.6512	100	100	564.7131	658.4129

Example 7.3. Let $H = L_2[0,1]$ and $C = \{p(t) \in L_2[0,1] \mid \int_0^1 tp(t)dt \leq 1\}$, and let $F = H \to H$ be the Volterra integral operator defined by $Fp(t) = \int_0^t p(s)ds$, $p(t) \in H$, $t \in [0,1]$.

Since F is $\frac{\pi}{2}$-inverse strongly monotone, we can choose $\tau_n = 0.25 \ \forall \ n \in \mathbb{N}$. Let α_n, β_n, γ_n, and θ_n be as in Example 7.4. The Cauchy error $\| J_n - J_{n-1} \| < 10^{-4}$ can define the stopping criterion. The different choices of J_0 and J_1 are given in Table 7.2 as follows as in Example 7.4. Then the results are presented in Table 7.2.

7.3.2 Convex Minimization Problem

Let a convex smooth function $A : H \to \mathbb{R}$ and a convex, lower-semicontinuous, and nonsmooth function $B : H \to \mathbb{R}$. The convex minimization problem (CMP)is to find $\hat{p} \in H \ni$

$$A(\hat{p}) + B(\hat{p}) \leq A(p) + B(p) \tag{7.41}$$

$\forall \ p \in H$. Problem (7.4) is equivalent, by Fermat's rule, to the problem of finding $\hat{p} \in H \ \ni$

$$0 \in \nabla A(\hat{p}) + \partial B(\hat{p}) \tag{7.42}$$

where ∇A represents the gradient of A and ∂B represents the subdifferential of B. In this part, we denote S for the minimizer of $A + B$. It is well known that if ∇A is $\frac{1}{L}$-Lipschitz continuous, then it is L-inverse strongly monotone ([28], Corollary 10). Moreover, ∂B is maximal monotone ([29], Theorem A). On substituting the values $F = \nabla A$ and $G = \partial B$ in Theorem 7.2, we have:

Theorem 7.4. *Let H be a real Hilbert space. Assume a convex and differentiable functiont $A : H \to \mathbb{R}$ with $\frac{1}{L}$-Lipschitz continuous gradient ∇A and a convex and lower semi-continuous function $B : H \to \mathbb{R}$ which $A + B$ attains a minimizer $\ni$ $S \neq \varnothing$. Let a real sequence $\{J_n\}$ defined by $C_1 = H$, $J_0, J_1 \in H$, and*

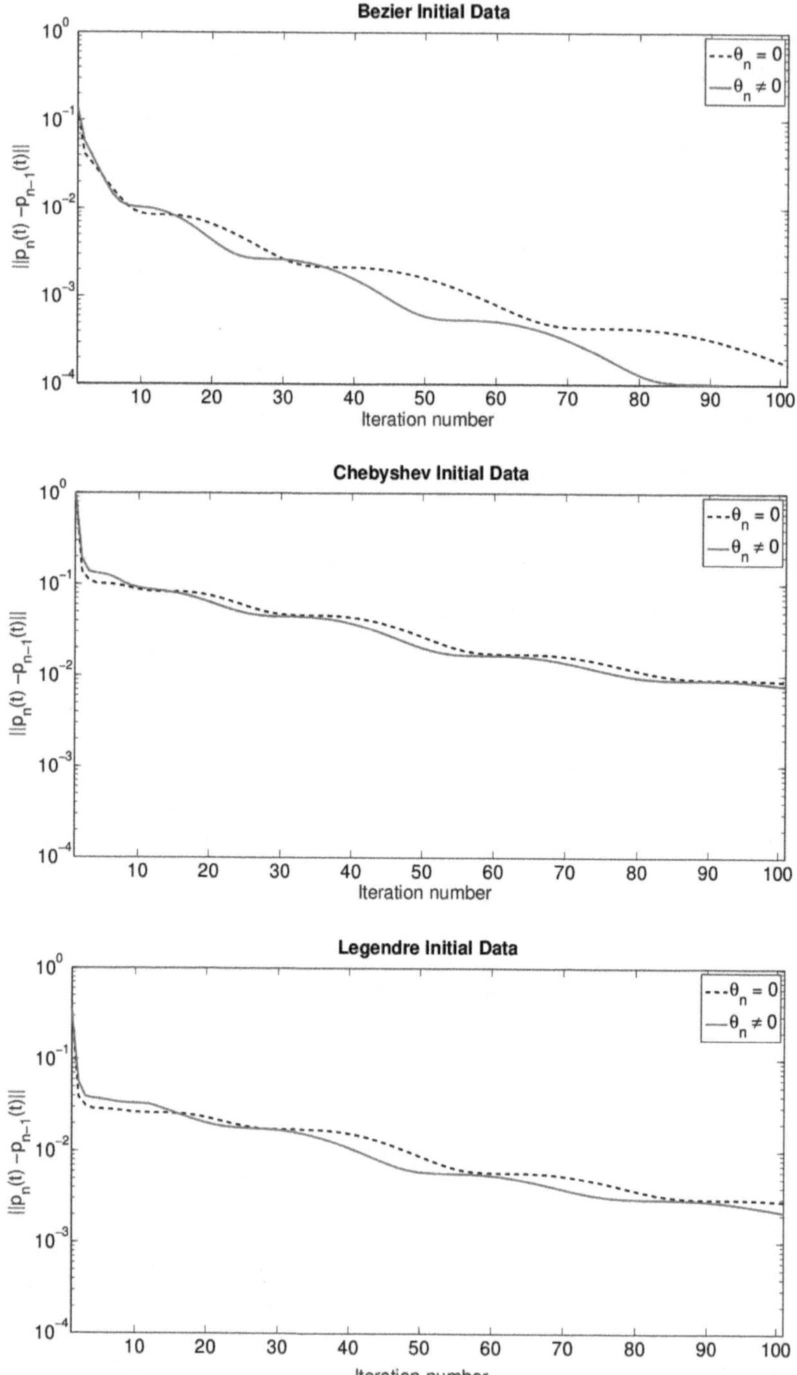

FIGURES 7.7–7.9 The error plotting $\| J_n - J_{n-1} \|$ with $\theta_n \neq 0$ and $\theta_n = 0$ for the variational inequality problem with choices $1-3$ for initial data of Iteration (3.2) from Table 7.2.

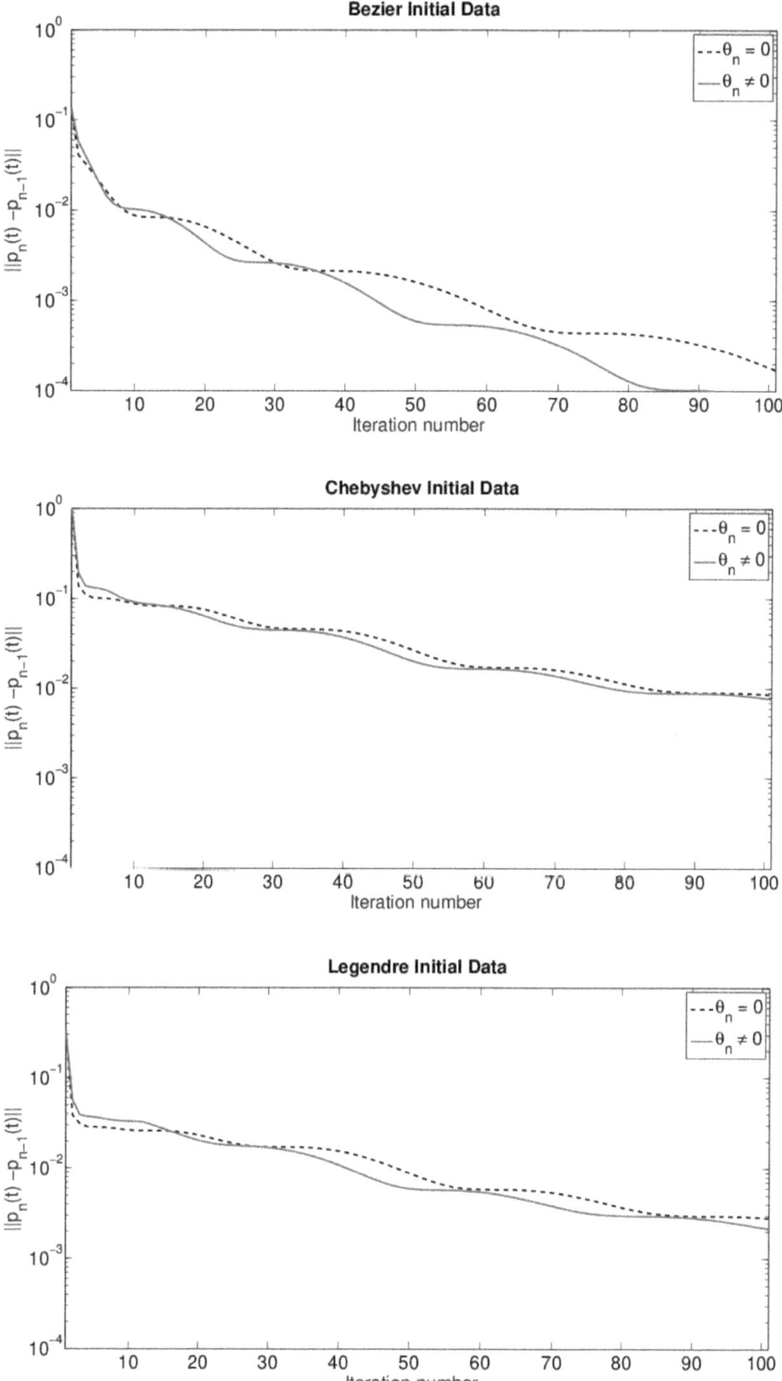

FIGURES 7.10–7.12 The error plotting $\| J_n - J_{n-1} \|$ with $\theta_n \neq 0$ and $\theta_n = 0$ for the variational inequality problem with choices $1-3$ for initial data of Iteration (3.3) from Table 7.2.

$$\begin{cases} \ell_n &= J_n + \theta_n (J_n - J_{n-1}) \\ r_n &= \gamma_n \ell_n + (1-\gamma_n) \Im_{\tau_n}^{\partial B} (\ell_n - \tau_n \nabla A(\ell_n)) \\ s_n &= \beta_n r_n + (1-\beta_n) \Im_{\tau_n}^{\partial B} (r_n - \tau_n \nabla A(r_n)) \\ t_n &= \alpha_n s_n + (1-\alpha_n) \Im_{\tau_n}^{\partial B} (u_n - \tau_n \nabla A(s_n)) \\ C_{n+1} &= \{ z \in C_n : \| t_n - z \|^2 \le \| J_n - z \|^2 + 2\theta_n^2 \| J_n - J_{n-1} \|^2 - 2\theta_n \langle J_n - z, J_{n-1} - x_n \rangle \} \\ J_{n+1} &= P_{C_{n+1}} J_1, \; n \ge 1 \end{cases} \tag{7.43}$$

where $\Im_{\tau_n}^{\partial B} = (I + \tau_n \partial B)^{-1}$, $\{\tau_n\} \subset (0, 2\alpha)$, $\{\theta_n\} \subset [0, \theta]$ for some $\theta \in [0,1)$ and the sequences $\{\alpha_n\}$, $\{\beta_n\}$, $\{\gamma_n\} \in [0,1]$. Assume that

(i) $\sum_{n=1}^{\infty} \theta_n \| J_n - J_{n-1} \| < \infty$;

(ii) $\limsup J_{n \to \infty} \gamma_n < 1$;

(iii) $0 < \liminf_{n \to \infty} \tau_n \le \limsup J_{n \to \infty} \tau_n < 2 / L$.

Then the sequence $\{J_n\}$ *converges strongly to* $\hat{J} = P_S J_1$.

Theorem 7.5. *Let* H *be a real Hilbert space. Assume that* $A : H \to \mathbb{R}$ *is a convex and differentiable function with* $\frac{1}{L}$*-Lipschitz continuous gradient* ∇A *and* $B : H \to \mathbb{R}$ *a convex and lower semi-continuous function which* $A + B$ *attains a minimizer* $\ni S \ne \varnothing$. *Let* $\{J_n\}$ *be a sequence defined by* $C_1 = H$, $J_0, J_1 \in H$, *and*

$$\begin{cases} \ell_n &= J_n + \theta_n (J_n - J_{n-1}) \\ r_n &= \gamma_n \ell_n + (1-\gamma_n) \Im_{\tau_n}^{\partial B} (\ell_n - \tau_n \nabla A(\ell_n)) \\ s_n &= \beta_n r_n + (1-\beta_n) \Im_{\tau_n}^{\partial B} (\ell_n - \tau_n \nabla A(r_n)) \\ t_n &= \alpha_n s_n + (1-\alpha_n) \Im_{\tau_n}^{\partial B} (s_n - \tau_n \nabla A(s_n)) \\ C_n &= \{ z \in H : \| t_n - z \|^2 \le \|^2 J_n - z \|^2 + 2\theta_n^2 \| J_n - J_{n-1} \|^2 - 2\theta_n \langle J_n - z, J_{n-1} - J_n \rangle \} \\ \ell_n &= \{ z \in \ell_{n-1} : \langle J_1 - J_n, J_n - z \rangle \ge 0 \} \\ J_{n+1} &= P_{C_n \cap \ell_n} J_1, n \ge 1 \end{cases} \tag{7.44}$$

where $\Im_{\tau_n}^{\partial B} = (I + \tau_n \partial B)^{-1}$, $\{\tau_n\} \subset (0, 2\alpha)$, $\{\theta_n\} \subset [0, \theta]$ *for some* $\theta \in [0,1)$ *and sequences* $\{\alpha_n\}$, $\{\beta_n\}$, $\{\gamma_n\} \in [0,1]$. *Assume that*

(i) $\sum_{n=1}^{\infty}\theta_n \| J_n - J_{n-1} \| < \infty;$

(ii) $\limsup J_{n\to\infty}\gamma_n < 1;$

(iii) $0 < \liminf_{n\to\infty}\tau_n \le \limsup J_{n\to\infty}\tau_n < 2/L.$

Then the sequence $\{J_n\}$ *converges strongly to* $\hat{J} = P_S J_1$.

Example 7.6. Solve the following minimization problem:

$$\min_{p(t)\in L_2[0,1]}\left\{2\| p(t)\|^2 + \left(1-t^2\right)p(t)\right\} \tag{7.45}$$

where $p(t)\in L_2[0,1]$, $t\in[0,1]$.

Set $A(p)=2\| p(t)\|^2 + \left(1-t^2\right)p(t)$ and $B(p)=0$ for all $p(t)\in L_2[0,1]$, $t\in[0,1]$. We have for $p\in L_2[0,1]$ and $\tau>0$, $\nabla A = 4p(t)+\left(1-t^2\right)$ and $\mathfrak{J}_{\tau}^{\partial B}(p)=I$, I is an identity. We conclude that ∇A is 4-Lipschitz continuous; therefore, it is $1/4$-inverse strongly monotone. Let $\tau_n = 0.2\ \ \forall\ \ n\in\mathbb{N}$. Assume that α_n, β_n, γ_n, and θ_n are as in Example 7.4. The stopping criterion is defined by $\| J_n - J_{n-1} \| < 10^{-4}$. The different choices of J_0 and J_1 are given in Table 7.3 as follows as in Example 7.4. Then the results are presented in Table 7.3.

According to the preliminary results obtained, the inertial three-step forward-backward splitting algorithms with the inertial technical term have a faster convergence rate than the standard three-step forward-backward splitting algorithms ($\theta n = 0$) for each of the choices.

Table 7.3
Comparison of $\theta_n \neq 0$ and $\theta_n = 0$ in Example 7.6

	Iteration (3.6)				Iteration (3.7)			
Choice of	Iteration Number		CPU Time (sec)		Iteration Number		CPU Time (sec)	
Initial Data	$\theta_n = 0$	$\theta_n \neq 0$	$\theta_n = 0$	$\theta_n \neq 0$	$\theta_n = 0$	$\theta_n \neq 0$	$\theta_n = 0$	$\theta_n \neq 0$
1	12	10	6.426603	4.932821	12	10	6.764630	6.559125
2	14	11	6.884614	5.538375	14	11	8.035529	7.211842
3	13	11	6.057024	5.922604	13	11	7.503243	7.119492

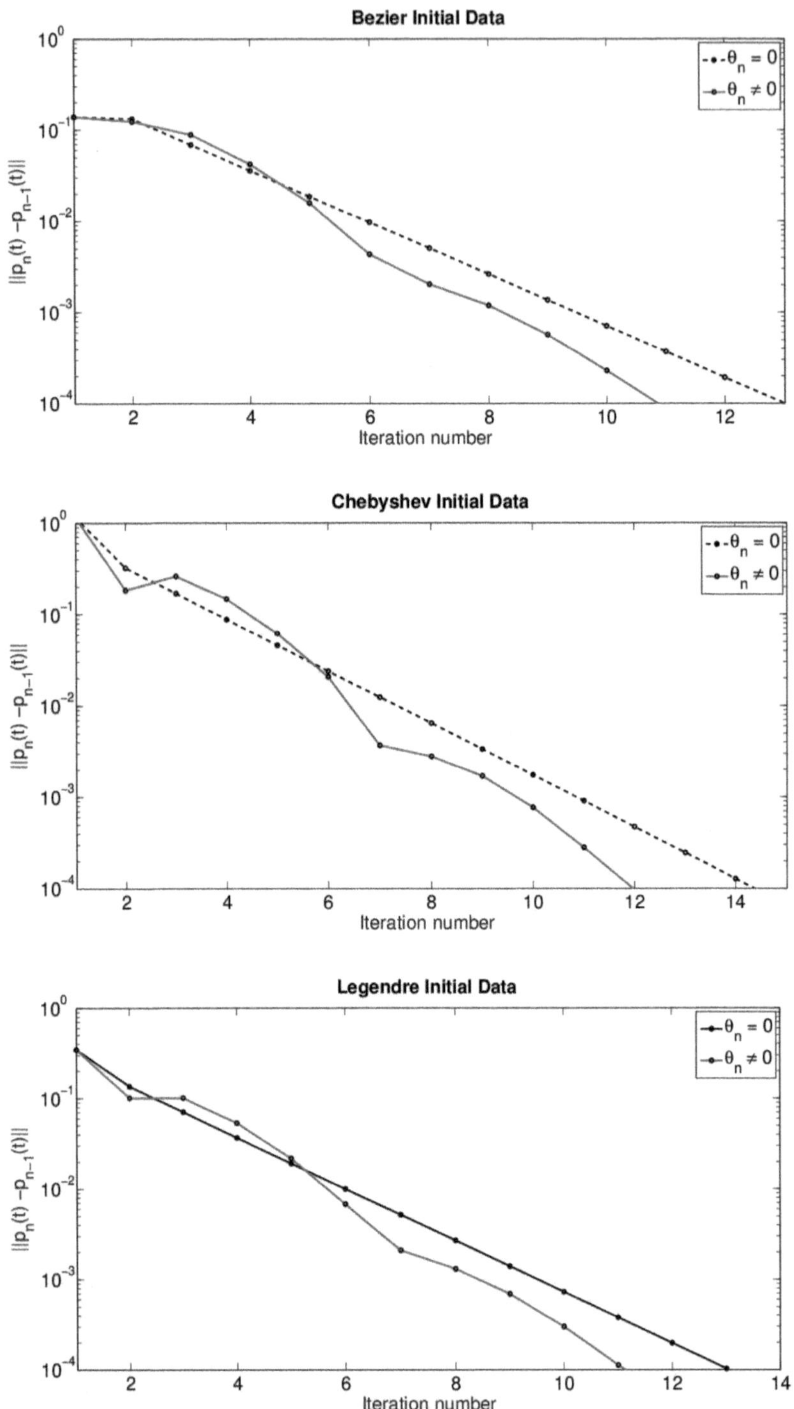

FIGURES 7.13–7.15 The error plotting $\| J_n - J_{n-1} \|$ with $\theta_n \neq 0$ and $\theta_n = 0$ for the convex minimization problem with choices $1-3$ for initial data of Iteration (3.6) from Table 7.3.

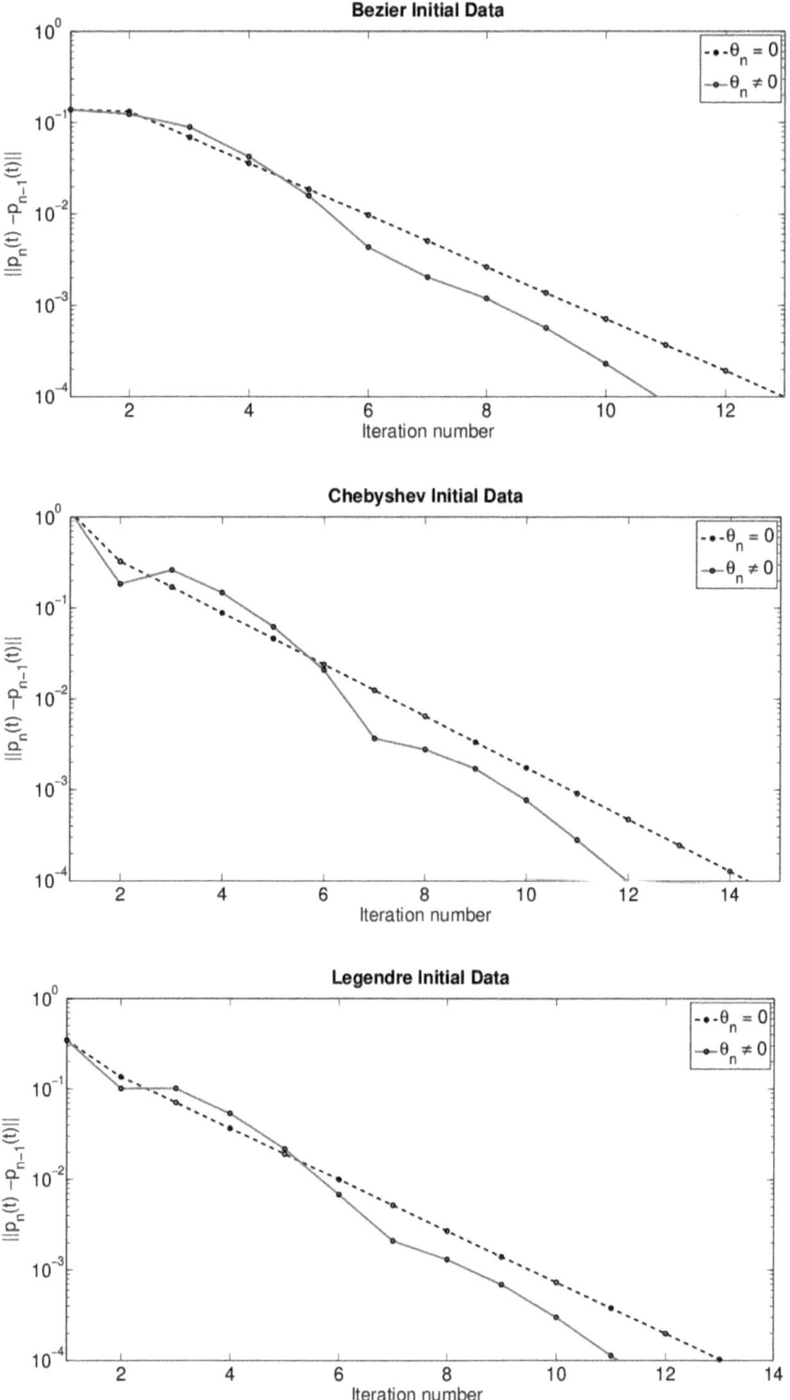

FIGURES 7.16–7.18 The error plotting $\| J_n - J_{n-1} \|$ with $\theta_n \neq 0$ and $\theta_n = 0$ for the convex minimization problem with choices $1-3$ for initial data of Iteration (3.7) from Table 7.3.

7.4 APPLICATION TO IMAGE RESTORATION PROBLEMS

The following linear equation system can solve the picture restoration problem in a one-dimensional vector:

$$b = Ax + \upsilon \tag{7.46}$$

where $x \in \mathbb{R}^{n\times 1}$ is an original image, and $b \in \mathbb{R}^{m\times 1}$ is the unknown image which is blurred by filter matrix $A \in \mathbb{R}^{m\times n}$ and added by the additive noise. In order to find solution for problem (7.46), we minimize the additive noise by the least squares (LS) problem, which is as follows

$$\min_{p} \frac{1}{2} \| b - Ap \|^2 \tag{7.47}$$

where $\|\cdot\|$ is the second norm defined by $\| p \| = \sqrt{\sum_{i=1}^{n} |J_i|^2}$. The solution of (7.47) can be estimated by many iterations.

In this section, we study the least squares problem (7.47) where $A : \mathbb{R}^n \to \mathbb{R}^n$ is a bounded linear operator and $b \in \mathbb{R}^n$ is fixed. We concluded that the inclusion problem (7.1) includes as a special case the LS problem (7.47), formulated with $F(p) = A^T(Ap - b)$ and $G(p) = p$. It is well known that $A^T(Ax - b)$ is Lipschitz continuous with Lipschitz constant $L = \| A^T A \|$. The image deblurring algorithm is applied. We know that the weak convergence sequence became the strong convergence sequence in a Euclidian space $\mathbb{R}^n$ where $n \in \mathbb{N}$. As a result, we shall employ our algorithm (7.1) in Theorem 7.1 to solve the unconstrained image restoration problem (7.47) because it's easy to use. Moreover, a variety of unconstrained iterative methods for image deblurring problems consisting of the OFBS(7.2), DR(7.3), FBS(7.6), and NV(7.9) algorithms are compared with our algorithm (7.1). The infinity norm is used to set the stopping criterion $\| p - J_n \|_\infty < 10^{-5}$, where p is the original image. The following parameters are used for our algorithm:

$$\tau_n = \frac{1}{\| A \|_1 \| A \|_\infty}, \theta = \frac{1}{4}, \alpha_n = \frac{n}{100n+1}$$

and $\alpha_n = \beta_n = \gamma_n$. We choose $\tau = \frac{1}{\| A \|_1 \| A \|_\infty}$ for OFBS(7.2) and DR(7.3); $\tau = \frac{1}{\| A \|_1 \| A \|_\infty}$, $\alpha_n = \frac{n}{100n+1}$ for FBS(7.6); and $\alpha_n = \frac{1}{100n}$, $\beta_n = \frac{n}{100n+1}$, $\tau_n = \frac{1}{\| A \|_1 \| A \|_\infty}$ for NV(7.9). We will present the restoration of images that have been corrupted by a Gaussian blur and motion blur. All algorithms are tested with degraded grey images by a Gaussian blur of filter size 20×20 with standard deviation $\sigma = 2$ and grey image by a motion blur specifying motion length 21 pixels (len = 21) and motion orientation 11^0 ($\Theta = 11$). The results are shown in the following figures.

FIGURES 7.19–7.30 The different original grey images: the degraded image by Gaussian blur of filter size 20×20 and $\sigma = 2$; the reconstructed images being used our iteration (2.1); and the comparison of the convergence history being used for the OFBS(1.2), DR(1.3), FBS(1.6), and NV(1.9) algorithms, respectively.

FIGURES 7.19–7.30 (Continued)

10000th Iteration

10000th Iteration

10000th Iteration

FIGURES 7.19–7.30 (Continued)

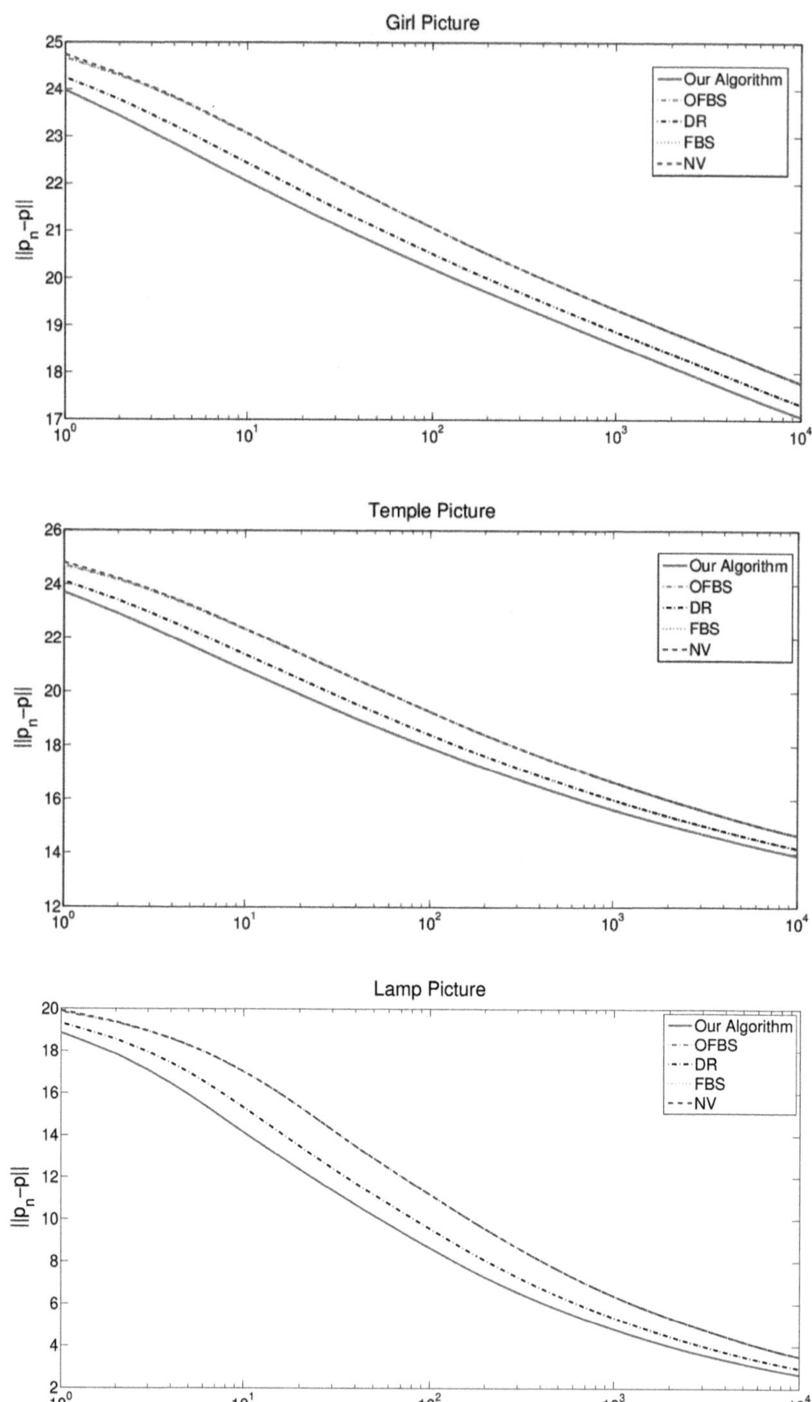

FIGURES 7.19–7.30 (Continued)

FIGURES 7.31–7.42 The different original grey images; the degraded image by motion blur with len 21 and $\Theta = 11$; the reconstructed images being used in our algorithm (7.1); and the comparison of the convergence history being used for the Richardson OFBS(1.2), DR(1.3), FBS(1.6), and NV(1.9) algorithms, respectively.

FIGURES 7.31–7.42 (Continued)

10000th Iteration

10000th Iteration

10000th Iteration

FIGURES 7.31–7.42 (Continued)

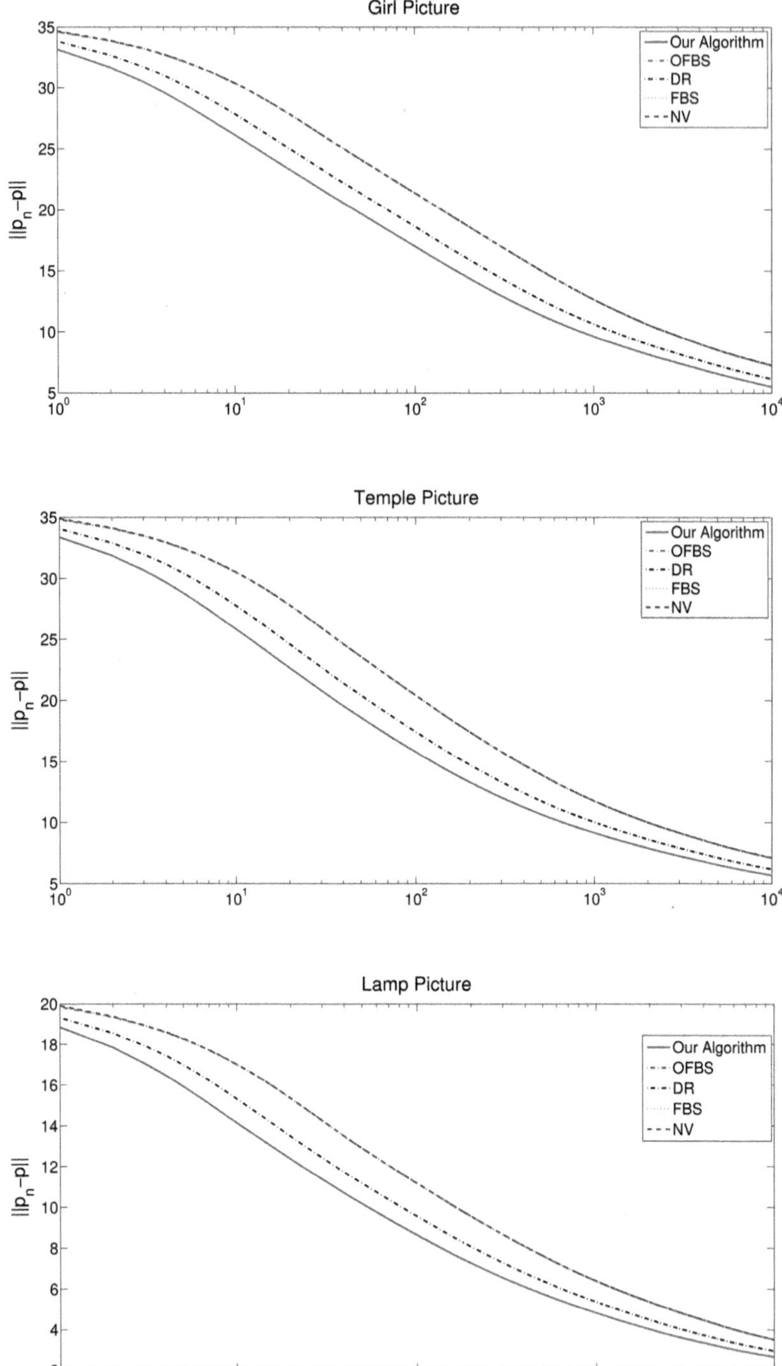

FIGURES 7.31–7.42 (Continued)

Next, we also apply our algorithm in solving the color deblurring problem. The following grey images illustrate an example of restoration of the degraded grey image by motion blur with len 21 and $\Theta = 11$.

From Figures 7.28 and 7.40–7.30–7.41, we see that our algorithm has a better quality of reconstructed images than the other four algorithms.

7.5 CONCLUSION

In this chapter, we present an inertial three-step forward-backward splitting algorithm for solving the inclusion problem. We prove the weak convergence theorem using some suitable conditions in Hilbert space. We also prove two strongly convergent theorems of our two algorithms constructed by two hybrid projection methods. Moreover, we apply our results to solve the variational inequality problem and the convex minimization problem. Some numerical experiments show that our inertial three-step forward-backward splitting algorithms have a competitive advantage over the standard three-step forward-backward splitting algorithms (see Tables 7.1–7.3 and Figures 7.1–7.18). Moreover, we applied our algorithms for solving unconstrained image recovery problems and compared our main algorithms with a viscosity forward-backward algorithm, original forward-backward algorithms, Douglas-Rachford, and forward-backward splitting algorithms. Our algorithms improved the reconstructed image quality compared with the other four unconstrained methods (see Figures 7.19–7.42).

7.6 ACKNOWLEDGMENTS

W. Cholamjiak and D. Yambangwai would like to thank the University of Phayao.

REFERENCES

[1] W. Takahashi, *Nonlinear Functional Analysis*, Yokohama Publishers, Yokohama, 2000.

[2] W. Takahashi, Fixed point theorems for new nonlinear mappings in a Hilbert space, *J. Nonlinear Convex Anal.* 11, 79–88 (2010).

[3] H. H. Bauschke and P. L. Combettes, *Convex Analysis and Monotone Operator Theory in Hilbert Spaces*, CMS Books in Mathematics, Springer, New York (2011).

[4] P. Cholamjiak, A generalized forward-backward splitting method for solving quasi inclusion problems in Banach spaces, *Numer. Algor.* 8, 221–239 (1994).

[5] P. L. Combettes and V. R. Wajs, Signal recovery by proximal forward-backward splitting, *Multiscale Model. Simul.* 4, 1168–1200 (2005).

[6] G. López, V. Martín-Márquez, F. Wang and H.K. Xu, Forward-Backward splitting methods for accretive operators in Banach spaces, *Abstr. Appl. Anal.* 2012, Art ID 109236 (2012).

[7] D. Lorenz and T. Pock, An Inertial forward-backward algorithm for monotone inclusions, *J. Math. Imaging Vision.* 51, 311–325 (2015).

[8] G. B. Passty, Ergodic convergence to a zero of the sum of monotone operators in Hilbert space, *J. Math. Anal. Appl.* 72, 383–390 (1979).

[9] P. Tseng, A modified forward-backward splitting method for maximal monotone mappings, *SIAM J. Control. Optim.* 38, 431–446 (2000).

[10] P. L. Lions and B. Mercier, Splitting algorithms for the sum of two nonlinear operators, *SIAM J. Numer. Anal.* 16, 964–979 (1979).

[11] J. Douglas and H. H. Rachford, On the numerical solution of the heat conduction problem in 2 and 3 space variables, *Trans. Amer. Math. Soc.* 82, 421–439 (1956).

[12] H. H. Bauschke and P. L. Combettes, A weak-to-strong convergence principle for Fejér-monotone methods in Hilbert spaces, *Math. Oper. Res.* 26, 248–264 (2001).

[13] D. Kitkuan, P. Kumam, A. Padcharoen, W. Kumam and P. Thounthong, Algorithms for zeros of two accretive operators for solving convex minimization problems and its application to image restoration problems. *J. Comput. Appl. Math.* 354, 471–495 (2018).

[14] W. Takahashi, Y. Takeuchi and R. Kubota, Strong convergence theorems by hybrid methods for families of nonexpansive mappings in Hilbert spaces, *J. Math. Anal. Appl.* 341, 276–286 (2008).

[15] K. Nakajo and W. Takahashi, Strong convergence theorems for nonexpansive mappings and nonexpansive semigroups, *J. Math. Anal. Appl.* 279(2), 372–379 (2003).

[16] B. T. Polyak, *Introduction to Optimization*, Optimization Software, New York (1987).

[17] B. T. Polyak, Some methods of speeding up the convergence of iterative methods, *Zh. Vychisl. Mat. Mat. Fiz.* 4, 1–17 (1964).

[18] F. Alvarez and H. Attouch, An inertial proximal method for monotone operators via discretization of a nonlinear oscillator with damping, *Set-Valued Anal.* 9, 3–11 (2001).

[19] F. Alvarez, Weak convergence of a relaxed and inertial hybrid projection-proximal point algorithm for maximal monotone operators in Hilbert spaces, *SIAM J. Optim.* 14(3), 773–782 (2004).

[20] Y. Dang, J. Sun and H. Xu, Inertial accelerated algorithms for solving a split feasibility problem, *J. Ind. Manag. Optim.* http://doi.org/103934/jimo.2016078.

[21] Q. L. Dong, H. B. Yuan, Y. J. Cho and Th. M. Rassias, Modified inertial Mann algorithm and inertial CQ-algorithm for nonexpansive mappings, *Optim. Lett.* http://doi.org/10.1007/s11590-016-1102-9.

[22] Y. Nesterov, A method for solving the convex programming problem with convergence rate $O\left(1/k^2\right)$, *Dokl. Akad. Nauk SSSR.* 269, 543–547 (1983).

[23] A. Moudafi and M. Oliny, Convergence of a splitting inertial proximal method for monotone operators, *J. Comput. Appl. Math.* 155, 447–454 (2003).

[24] F. E. Browder, Nonexpansive nonlinear operators in a Banach space, *Proc. Natl. Acad. Sci. USA.* 54, 1041–1044 (1965).

[25] S. Suantai, Weak and strong convergence criteria of Noor iterations for asymptotically nonexpansive mappings, *J. Math. Anal. Appl.* 311, 506–517 (2005).

[26] C. Martinez-Yanes and H.-K. Xu, Strong convergence of CQ method for fixed point iteration processes. *Nonlinear Anal.* 64, 2400–2411 (2006).

[27] R. Glowinski, J.-L. Lions and R. Tremolieres, *Numerical Analysis of Variational Inequalities*, North-Holland, Amsterdam (1981).

[28] J. B. Baillon and G. Haddad, Quelques proprietes des operateurs angle-bornes et cycliquement monotones, *Isr. J. Math.* 26, 137–150 (1977).

[29] R. T. Rockafellar, On the maximality of subdifferential mappings, *Pac. J. Math.* 33, 209–216 (1970).

8 Explicit Study of Fractal Generation of Mandelbrot and Julia Sets via Picard–*S* Iteration Scheme with *S*–Convexity

Nisha Sharma, Hemen Dutta, and Shikha Pandey

8.1 INTRODUCTION

Complex graphics of nonlinear dynamical systems is a captivating field of research with innumerable contributions to science, art and textile manufacturing, technology, and several other fields of human endeavor. The Mandelbrot set is the most well-known endlessly complex object in Benoit Mandelbrot's fractal theory [1]. Researchers have been analyzing this object substantially since its inception [1–3]. Various authors [4–14] have investigated the Mandelbrot set and its numerous generalizations; specifically, the graphics of Mandelbrot sets and Julia sets studied by Kwun et al. inspired us to study the automatic aesthetic patterns of the Picard–*S* iteration scheme. External and internal disruptions of the Mandelbrot set have been analyzed in [15], and this object has been investigated from several perspectives. With the use of the Mann iteration approach, Rani and Kumar [16] developed improved Mandelbrot sets. Using the Ishikawa iteration scheme, Rana et al. [17] established comparative superior Mandelbrot sets. They also considered relatively superior Mandelbrot sets of quadratic, cubic, and other complex value polynomials and discussed some characteristics. In 1918, Gaston Julia [18] attempted to obtain the iteration procedure of the complex function

$$f(z) = z^2 + cf(z) = z^2 + c$$

where c is a complex number and generates the Julia set. Julia sets are phenomenal illustrations of computational research that were revolutionary at their time. When computer graphics became accessible, these mathematical objects were seen [19]. Benoit Mandelbrot, on the other hand, introduced the Mandelbrot set in 1979 using

DOI: 10.1201/9781003322856-8

c as a complex measure in a quadratic function which was complex [20]. Mandelbrot and Julia sets are two of the most well-known instances of highly complex chaotic systems generated by a very basic mathematical procedure. Mandelbrot introduced the term "fractal" to characterize some rather self-similar patterns. Fractals are so much more than complicated shapes and photorealistic images generated by computers. A fractal can be anything that appears random and anomalous. Fractals are the distinctive, random patterns left behind by the chaotic world's unforeseeable mobility at work. Fractal image compression is among the most widely used technology of fractals in computer science. This type of compression makes use of the fact that fractal geometry accurately describes the real world. Images are compressed much more effectively in this manner than in traditional methods (e.g., JPEG or GIF file formats). Another benefit of fractal compression is that there is no pixelization when the image is enlarged. When the image's size is increased, it often appears to be better. Superior Julia sets and Mandelbrot sets were described in 2004 by Rani and Kumar [16] using the Mann iteration scheme, which is a one-step iterative approach. Rana et al. [17, 21] proposed relative superior Julia sets and Mandelbrot sets using the Ishikawa iteration framework, which is a two-step fixed point effective algorithm.

Kang et al. [22, 23] introduced relatively good Mandelbrot quantities and tricorn multicorns through the s–iteration scheme. In addition, how to generate fractal images of the cytolation method was discussed, and it was proved that the cytolation scheme converges faster at a more complex level than the Ishikawa iteration scheme. The Julia set and Mandelbrot set using the Noor iterative process, which is a three-step iterative procedure, are shown in [24]. Convexity, approximation convexity, and generalization of results by Bernstein and Doetsch [25] are dealt with in the work presented. Breckner [26] introduced the concept of convexity and rational convexity. Some new results on the Hadamard inequality for convex functions are explained in [27–29]. Hudzik and Maligranda [30] discussed some results related to convex functions in the second sense. Takahashi [31] established the notion of convex metric space, which is a more general space, and each linear normed space is a special case of it. Numerous papers explored the intersection of an s–convex combination [32] and various iterative process. Mishra et al. [33, 34] used Ishikawa iteration with s–convexity to obtain fixed point results for relative superior Julia sets, tricorns, and multicorns. Kwun et al. [35, 36] recently generated fractals with s–convexity incorporating Jungck-CR and modified Jungck S iterations. The Picard–S iteration approach is used in this chapter to enhance fractals. Other Mandelbrot sets in the literature are different from our observations.

The following is how the chapter is structured: We begin with some fundamental precepts in Section 8.2. Section 8.3 deals with the fractal generation in the framework of the Picard–S iteration scheme. The escape criterion for cubic polynomials is established in Section 8.4. The escape criterion for complex quadratic polynomials is established in Section 8.5. Furthermore, the escape criterion for higher-degree polynomials is established in Section 8.6. In this section, we also present generalized Mandelbrot sets and Julia sets. Eventually, finally, we make some final observations.

8.2 PRELIMINARIES

In 2007, Agarwal et al. [37] addressed the following question: Is it possible to develop an iteration process whose rate of convergence is faster than the Picard iteration [38]? They answered this problem by introducing an $S-$iteration [39, 40] method. It was shown in [39, 40] that the $S-$iteration method converges at a rate the same as that of the Picard iteration method and faster than the Mann iteration method [41] for the class of contraction mappings. In 2009, Sahu [40] was interested in the same problem and proved both theoretically and numerically that the $S-$iteration method converges at a rate faster than both the Picard iteration method and the Mann iteration method for the class of contraction mappings. A recent study by Chugh et al. [42] shows that the CR iterative method [42] converges faster than the Picard, Mann, Ishikawa [43], Noor [44], Phuengrattana and Suantai [45], and *Sahu* [39, 40] iterative methods for a particular class of quasi-contractive operators, which include the aforementioned class of contraction operators. Inspired by these works, Gursoy and Karakaya propose the following problem:

Problem. Is it possible to develop an iteration process whose rate of convergence is even faster than the iteration [42]?

To answer this problem, Gursoy and Karakaya introduced the following iteration method called Picard–S iteration, which is as follows:

Let D be a nonempty convex subset of a Banach space B and T a self-map of D.

$$\begin{cases} x_0 \in D \\ x_{\eta+1} = Ty_\eta \\ y_\eta = \left(1-\alpha_\eta^1\right)Tx_\eta + \alpha_\eta^1 Tz_\eta \\ z_\eta = \left(1-\alpha_\eta^2\right)x_\eta + \alpha_\eta^2 Tx_\eta, \qquad \eta \in \mathbb{N} \end{cases}$$

In [46], it is shown that the Picard–S iteration method can be used to approximate fixed point of contraction mappings. Also, it is shown that Picard–S iteration is equivalent to and converges faster than the CR iteration method for contraction mappings. It was exemplified that the Picard–S iteration method converges faster than the CR iteration method and hence it is faster than all Picard, Mann, Ishikawa, Noor, PS, Sahu, and other iteration methods in the existing literature. A data dependence result was proved for fixed points of contraction mappings with the help of the new iteration method. Also, it was shown that the Picard–S iteration method can be used as an effective method to solve differential equations with retarded arguments.

Now, to make this chapter self-contained, the following definitions and lemmas are important and are as follows:

Definition 8.1. ([47], Julia set) Let $(\mathcal{J}; \mathcal{C})$ be a dynamical system on which the escape time algorithm can be applied, and $\mathcal{J}: \mathcal{C} \to \mathcal{C}$, where $\mathcal{C}$ symbolizes a polynomial of degree ≥ 1. Let $\mathcal{F}_\mathcal{J}$ be the set of points in $\mathcal{C}$ whose orbits do not converge to the point at ∞. That is,

$$F_J = \left\{\sigma \in C : \{\left|J^n(\sigma)\right|\}_{n=0}^{\infty} \text{ is bounded}\right\}$$

$\mathcal{F}_{\mathcal{J}}$ is called the filled Julia set of the polynomial $\mathcal{J}$. The boundary points of $\mathcal{F}_{\mathcal{J}}$ are called the points of the Julia set of the polynomial $\mathcal{J}$ or simply the Julia set.

Definition 8.2. ([48], Mandelbrot set). The Mandelbrot set $\mathcal{M}$ consists of all parameters c for which the Julia set of $\mathcal{Q}_c$ is connected; that is

$$\mathcal{M} = \{c \in \mathcal{C} : \mathcal{K}(\mathcal{Q}_c)\, is\ connected\} \tag{8.1}$$

In fact, $\mathcal{M}$ contains an enormous amount of information about the structure of Julia sets. The Mandelbrot set $\mathcal{M}$ for the quadratic $\mathcal{Q}_c\,(z) = z^2 + c$ is defined as the collection of all $c \in \mathcal{C}$ for which the orbit of the point 0 is bounded; that is

$$\mathcal{M} = \{c \in \mathcal{C} : \mathcal{Q}_c^n\,(0); n \in \mathcal{N}\,(set\ of\ all\ natural\ numbers)\, is\ bounded\} \tag{8.2}$$

We choose the initial point 0, as 0 is the only critical point of $\mathcal{Q}_c$.

Definition 8.3. Let $\mathcal{C}$ be a nonempty set and $\mathcal{J}$: $\mathcal{C} \to \mathcal{C}$ be a mapping. For any point $\hbar_0 \in$ C, the Picard orbit is defined as the set of iterates of a point $\hbar_0$; that is

$$O(\mathcal{J}, \hbar_0) = \{\hbar_n ; \hbar_n = \mathcal{J}(\hbar_{n-1}), n \in N\} \tag{8.3}$$

where $O(\mathcal{J}, \hbar_0)$ of $\mathcal{J}$ at the initial point $\hbar$ is the sequence $\{\mathcal{J}^n\,(\hbar_n)\}$.

Theorem 8.1. Let $\mathcal{Q}_c\,(\sigma) = \sigma\kappa + 1 + c$, where $\kappa = 1, 2, \ldots$ and $c \in \mathcal{C}$.

(1) (escape criterion for the Picard iteration [49]) Suppose that

$$|\sigma_0| > max\left\{|c|, 2^{\frac{1}{k}}\right\}$$

Then, for σ_n given by the Picard iteration, that is, $\sigma_n = \mathcal{Q}_c\,(\sigma_{n-1})$, we have $|\sigma_n| \to \infty$ as

$n \to \infty$.

(2) (escape criterion for the Mann iteration [16]) Suppose that

$$|\sigma_0| > max\left\{|c|, \left(\frac{2}{\alpha}\right)^{\frac{1}{k}}\right\}$$

where $\alpha \in (0, 1]$. Then, for σ_n given by the Mann iteration, that is, $\sigma_n = (1 - \alpha)\sigma_{n-1} + \alpha Qc\,(\sigma_{n-1})$, we have $|\sigma_n| \to \infty$ as $n \to \infty$.

Then, for σ_n given by the Mann iteration, that is, $\sigma_n = (1-\alpha)\sigma_{n-1} + \alpha\mathcal{Q}_c(\sigma_{n-1})$, we have $|\sigma_n| \to \infty$ as $n \to \infty$.

Definition 8.4. (Picard–Mann orbit [50]) Let $\mathcal{S} \subset \mathcal{C}$ and $\mathcal{J} : \mathcal{S} \to \mathcal{S}$. Consider a sequence σ_n of iterates for the initial point $\sigma_0 \in \mathcal{C}$ such that

$$\begin{cases} \sigma_{n+1} = \mathcal{J}\omega_n \\ \omega_n = (1-\alpha_n)\sigma_n + \alpha_n \mathcal{J}\sigma_n, \quad n = 0,1,2 \end{cases}$$

where $\eta_n \in 0,1$. This sequence of iterates is called the Picard–Mann orbit, which is a function of three arguments $(\mathcal{J},\sigma_0,\alpha_n)$ and will be denoted by $PMO(\mathcal{J},\sigma_0,\alpha_n)$.

Definition 8.5. (Picard–S orbit) Let $\mathcal{C}$ be a nonempty set and $\mathcal{J}:\mathcal{C}\to\mathcal{C}$ be a mapping. Consider a sequence $\hbar_n$ of iterates for initial point $\hbar_n \in \mathcal{C}$ such that

$$\begin{cases} \hbar_n &= \left(1-\eta_n^2\right)\sigma_n + \eta_n^2 \mathcal{J}\sigma_n; \\ \ell_n &= \left(1-\eta_n^1\right)\mathcal{J}\sigma_n + \eta_n^1 \mathcal{J}\hbar_n; \\ \sigma_{n+1} &= \mathcal{J}\ell_n; n \geq 0 \end{cases}$$

where η_n^1 and $\eta_n^2 \in (0,1]$, and $\{\eta_n^1\}$, $\{\eta_n^1\}$ are sequences of positive numbers. This sequence of iterates is called the Picard–S orbit, which is a function of four tuples $(\mathcal{J},\hbar_0,\eta_n^1,\eta_n^2)$.

Definition 8.6. (s –convex combination [31]) Let $\hbar_1,\hbar_2,\hbar_3,\ldots,\in\mathcal{C}$ and $s\in(0,1]$. The s –convexity combination is defined in the following way

$$\eta_1^s\hbar_1 + \eta_2^s\hbar_2 + \eta_3^s\hbar_3 + \ldots + \eta_1^s\hbar_n$$

where $\eta_k \geq 0$ for $k \in 1,2,3,\ldots,n$ and $\sum_{n=1}^{\infty}\eta_k = 1$.

It is noticed that for $s=1$, the s –convexity combination arranges to the normal convex combination. We shall write the s –convexity combination in the Picard–S iteration. We take $\hbar_0 - \hbar = \hbar \in \mathcal{C}, \eta_k^1 = \eta_1, \eta_k^2 = \eta_1$; then we can write the Picard–S iteration scheme with s – convexity in the following way, where $\mathcal{Q}_\xi(\hbar_n)$ is a quadratic, cubic, or $(k+1)$ degree polynomial.

$$\begin{cases} \hbar_n &= (1-\eta_2)^s\sigma_n + \eta_2^s\mathcal{Q}_\xi(\sigma_n); \\ \ell_n &= (1-\eta_1)^s\mathcal{Q}_\xi(\sigma_n) + \eta_1^s\mathcal{Q}_\xi(\hbar_n); \\ \sigma_{n+1} &= \mathcal{Q}_\xi(\ell_n); n \geq 0, n \geq 0 \end{cases}$$

where $\eta_1,\eta_2 \in (0,1]$.

Escape criteria perform an important role in the analysis and generation of Mandelbrot sets and Julia sets. Now, we define escape criteria for Mandelbrot sets and Julia sets in Picard–S orbit with s –convexity. It is important to note that that by using Picard–S with s –convexity, we will generate completely new orbits and, as a result, new fractal forms during the generation process.

8.3 FRACTAL GENERATION

In this section, we present some visual illustrations of fractal patterns (Mandelbrot and Julia sets) generated by Picard–S iteration with s –convexity. The escape time algorithms are used to generate these fractals. Figures 8.1–8.6 describe the

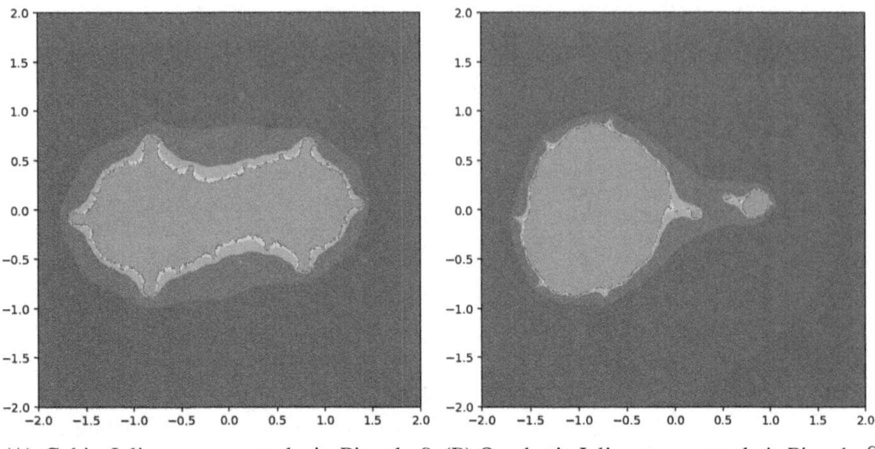

(A) Cubic Julia set generated via Picard–S iteration. (B) Quadratic Julia set generated via Picard–S iteration.

FIGURE 8.1 Cubic and quadratic Mandelbrot set generated via Picard–S iteration with $s-$convexity for $\eta_1 = 0.1, \eta_2 = 0, s = 1$.

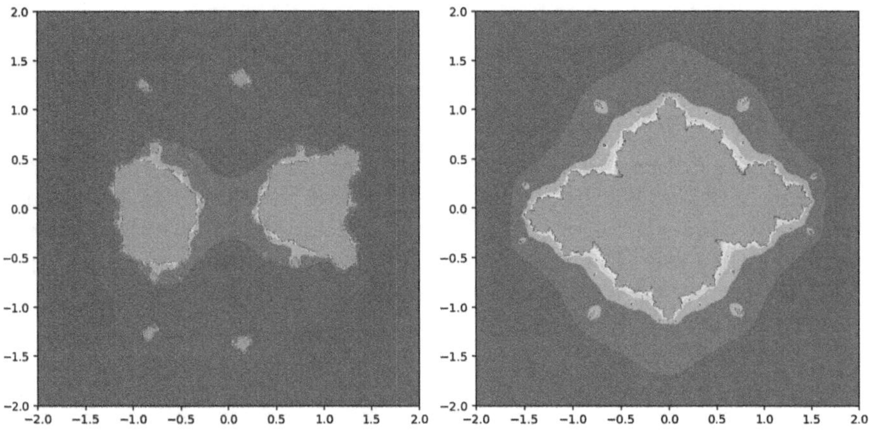

(A) Cubic Julia set generated via Picard–S iteration. (B) Quadratic Julia set generated via Picard–S iteration.

FIGURE 8.2 Cubic and quadratic Julia set generated via Picard–S iteration with $s-$convexity for $\eta_1 = 0.1, \eta_2 = 1, s = 1$.

Julia sets for various values of η_1, η_2, and s, while Figures 8.7–8.19 show the Mandelbrot sets. We used a different color map for Mandelbrot sets and all Julia sets. Julia sets and Mandelbrot sets are generated for quadratic and cubic functions using the escape time algorithm with the escape criterion attained in the previous section. Comprehensively, Figures 8.1(A) and 8.1(B) represent the cubic Julia set and quadratic Julia set generated via Picard–S iteration for $\eta_1 = 0.1$, $\eta_2 = 0$, and s.

Figures 8.2(A) and 8.2(B) show cubic and quadratic Julia sets generated via Picard S–iteration with $s-$convexity for $\eta_1 = 0.1$, $\eta_2 = 1$, and $s = 1$. Figure 8.3(B) shows cubic and quadratic Julia sets generated via Picard S–iteration with $s-$convexity $\eta_1 = 0.1$, $\eta_2 = 1$, and $s = 2$.

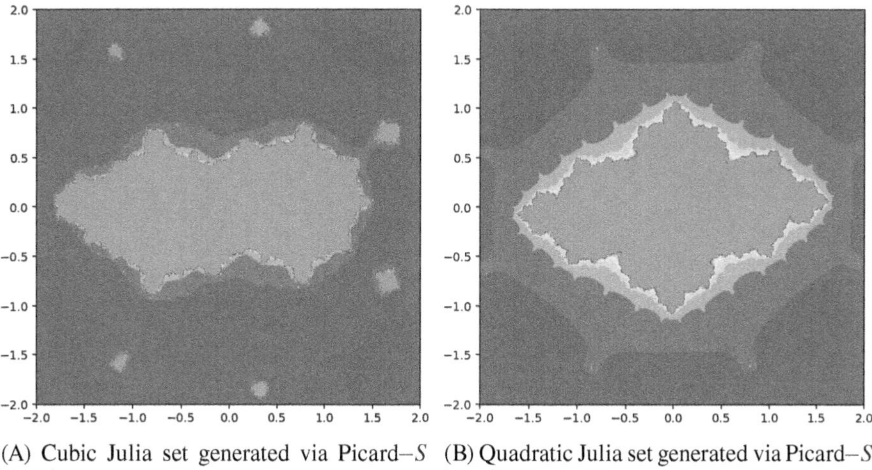

(A) Cubic Julia set generated via Picard–S iteration. (B) Quadratic Julia set generated via Picard–S iteration.

FIGURE 8.3 Cubic and quadratic Julia set generated via Picard–S iteration with s –convexity for $\eta_1 = 0.1, \eta_2 = 1, s = 2$

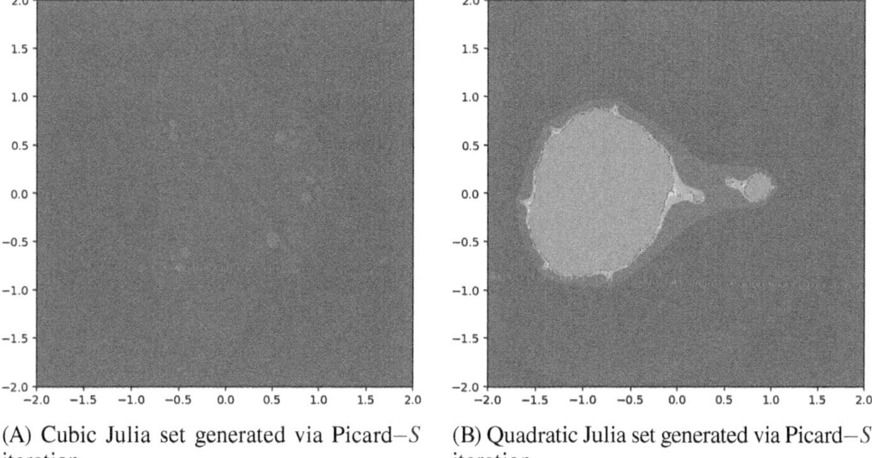

(A) Cubic Julia set generated via Picard–S iteration. (B) Quadratic Julia set generated via Picard–S iteration.

FIGURE 8.4 Cubic and quadratic Julia sets generated via Picard–S iteration with s –convexity for $\eta_1 = 1, \eta_2 = 0.1, s = 0$.

Figure 8.4(A) shows cubic and quadratic Julia sets generated via Picard–S iteration with s –convexity for $\eta_1 = 1$, $\eta_2 = 0.1$, and $s = 0$, in which the variation is clearly shown. Figure 8.5(A) shows cubic and quadratic Julia sets generated via Picard–S iteration with s –convexity for $\eta_1 = 1$, $\eta_2 = 0.1$, and $s = 1$, in which the variation is clearly shown. For different values of s and the same values for η_1 and η_2, the variation in cubic Julia sets and quadratic Julia sets can can be analyzed (refer to Figures 8.5 and 8.6(A)).

Figure 8.6(A) shows the clear variation in cubic and quadratic Julia sets generated via Picard–S iteration with s –convexity for $\eta_1 = 1$, $\eta_2 = 0.1$, and $s = 1$.

Figures 8.7–8.19 show the Mandelbrot sets for cubic and quadratic functions in which Figures 8.7(A), 8.7(B), 8.8(A), and 8.8(B) show the difference in fractals for distinct η_1 values and the same η_2 values.

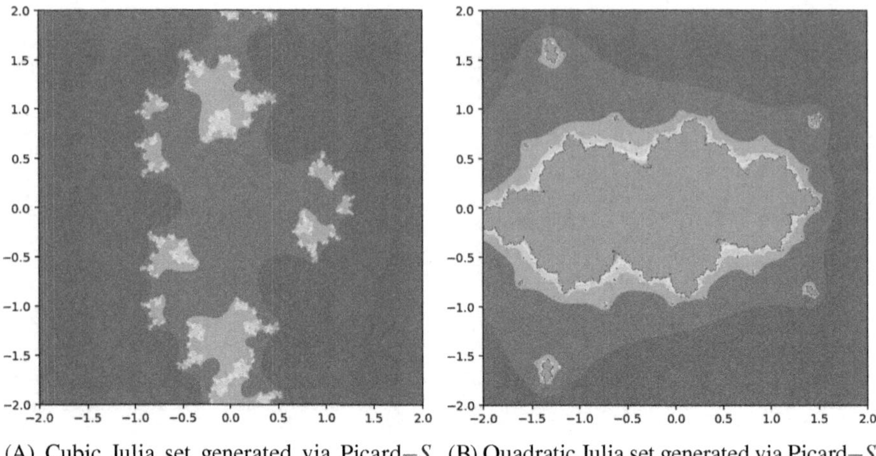

(A) Cubic Julia set generated via Picard–S iteration.

(B) Quadratic Julia set generated via Picard–S iteration.

FIGURE 8.5 Cubic and quadratic Julia sets generated via Picard–S iteration with s –convexity for $\eta_1 = 1, \eta_2 = 0.1, s = 1$.

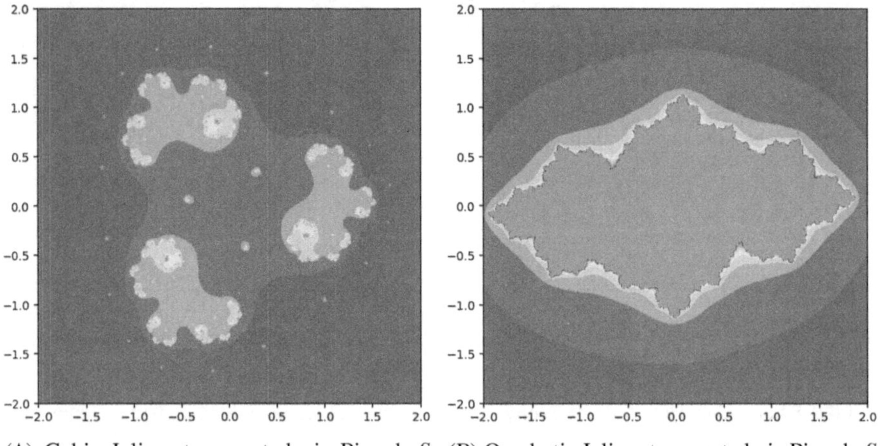

(A) Cubic Julia set generated via Picard–S iteration.

(B) Quadratic Julia set generated via Picard–S iteration.

FIGURE 8.6 Cubic and quadratic Julia sets generated via Picard–S iteration with s –convexity for $\eta_1 = 1, \eta_2 = 0.1, s = 2$.

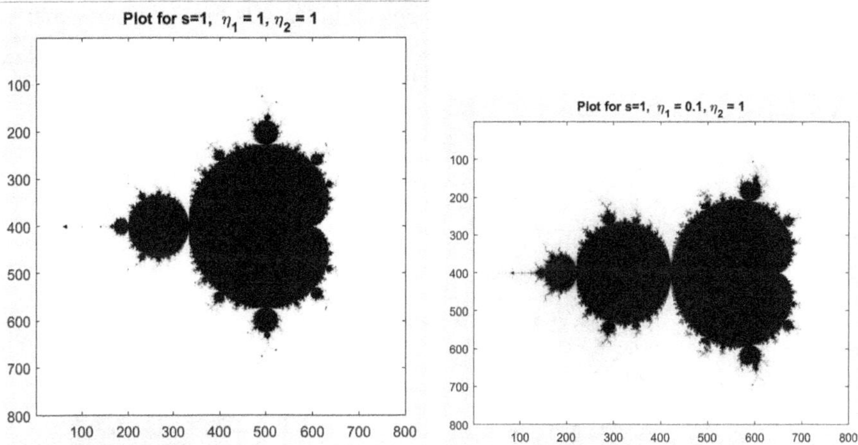

(A) Quadratic Mandelbrot set generated via Picard–S iteration with s–convexity.

(B) Quadratic Mandelbrot set generated via Picard–S iteration with s–convexity.

FIGURE 8.7 Quadratic Mandelbrot set generated via Picard–S iteration with s–convexity for $\eta_1 = 1$ and $\eta_1 = 0.1$ and with equal η_2.

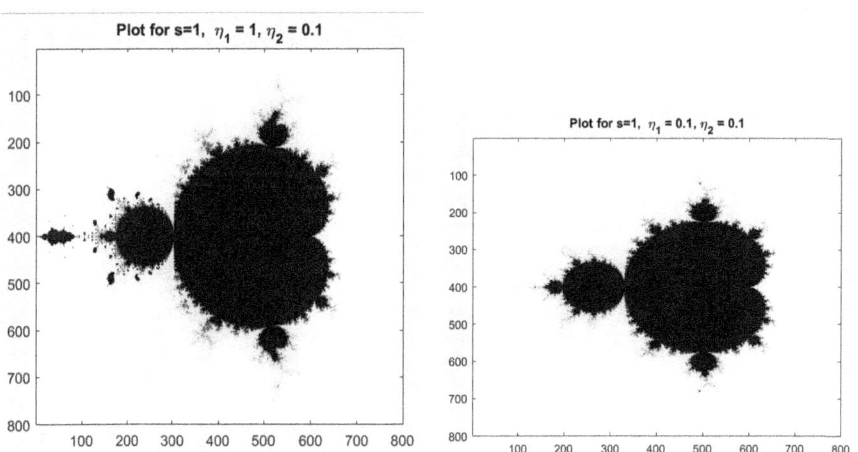

(A) Quadratic Mandelbrot set generated via Picard–S iteration with s–convexity.

(B) Quadratic Mandelbrot set generated via Picard–S iteration with s–convexity.

FIGURE 8.8 Quadratic Mandelbrot set generated via Picard–S iteration with s–convexity for $\eta_1 = 1, \eta_2 = 0.1$, and $\eta_1 = 0.1, \eta_2 = 0.1$.

Figures 8.9(A) and 8.9(B) show the difference in fractals for the same η_1 and η_2 values, where both $\eta_1 =_2= 0.1$, the fractal generation in this case, is quite same, whereas Figures 8.10(A) and 8.10(B) show the difference in fractals for same η_1 and η_2 values, where both $\eta_1 =_2= 1$.

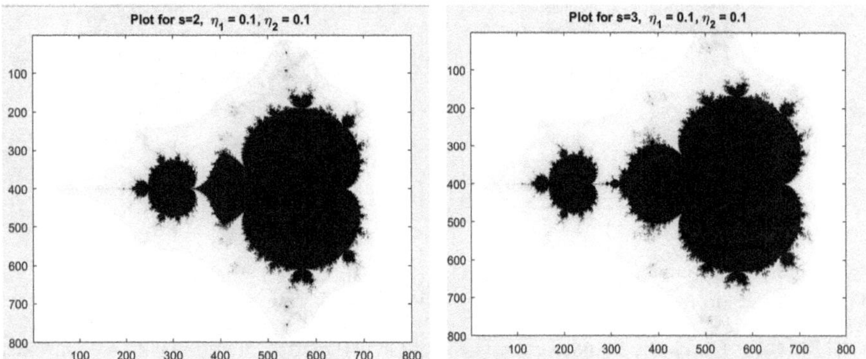

(A) Quadratic Mandelbrot set generated via Picard–S iteration with s –convexity.

(B) Quadratic Mandelbrot set generated via Picard–S iteration with s –convexity.

FIGURE 8.9 Quadratic Mandelbrot set generated via Picard–S iteration with s –convexity for equal η_i, $i = 1,2$ where $\eta_i = 0.1$.

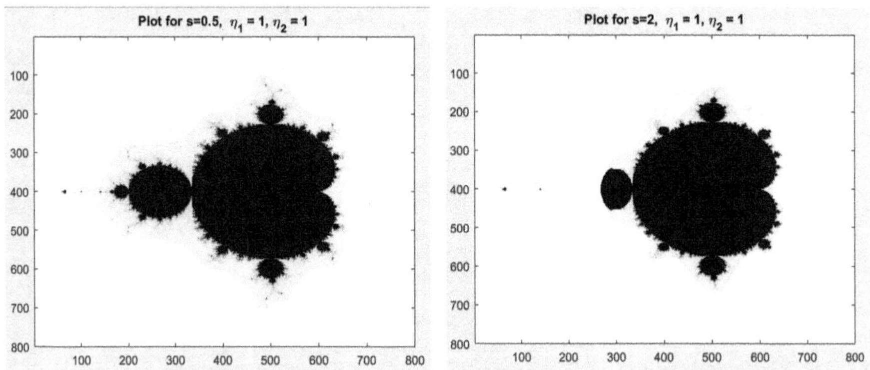

(A) Quadratic Mandelbrot set generated via Picard–S iteration with s –convexity.

(B) Quadratic Mandelbrot set generated via Picard–S iteration with s –convexity.

FIGURE 8.10 Quadratic Mandelbrot set generated via Picard–S iteration with s –convexity for equal η_i, $i = 1,2$ where $\eta_i = 1$.

The huge variation in Quadratic Mandelbrot set fractals can easily be seen in Figures 8.11(A) and 8.11(B) and 8.12(A) and 8.12(B), in which both η_1 and η_2 are distinct. For 11(A), $\eta_1 = 0.8, \eta_2 = 0.2$ and for 11(B), $\eta_1 = 0.2, \eta_2 = 0.8$, and for 12(A), $\eta_1 = 2, \eta_2 = 1$ and for 12(B), $\eta_1 = 1, \eta_2 = 2$.

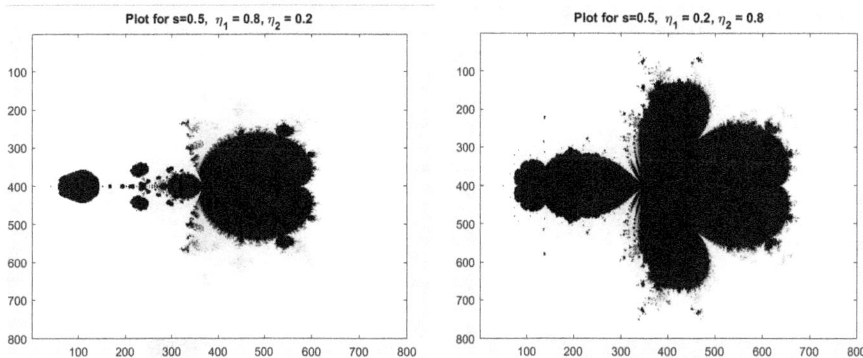

(A) Quadratic Mandelbrot set generated via Picard–S iteration with s –convexity.

(B) Quadratic Mandelbrot set generated via Picard–S iteration with s –convexity.

FIGURE 8.11 Quadratic Mandelbrot set generated via Picard–S iteration with s –convexity for $\eta_1 = 0.8, \eta_2 = 0.2$, and $\eta_1 = 0.2, \eta_2 = 0.8$.

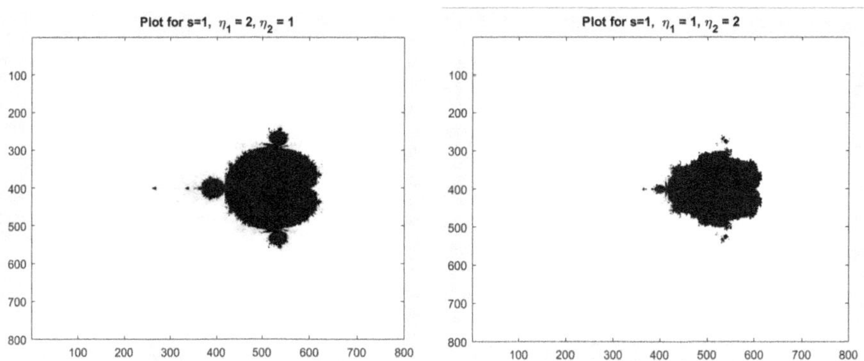

(A) Quadratic Mandelbrot set generated via Picard–S iteration with s –convexity.

(B) Quadratic Mandelbrot set generated via Picard–S iteration with s –convexity.

FIGURE 8.12 Quadratic Mandelbrot set generated via Picard–S iteration with s –convexity for $\eta_1 = 2, \eta_2 = 1$, and $\eta_1 = 1, \eta_2 = 2$.

Figures 8.13–8.19 show the fractals for the Mandelbrot sets for the cubic functions. Figures 8.13–1.19 shows the Mandelbrot sets for the cubic function in which Figures 8.13(A) and 8.13(B) show the difference in fractals for the same η_2 values and different η_1 values, whereas Figures 8.14(A) and 8.14(B) show the fractal pattern difference, where $\eta_1 = 1, \eta_2 = 0.1$ is taken for 8.14(A) and $\eta_1 = 1, \eta_2 = 0.1$ is taken for 8.14(B).

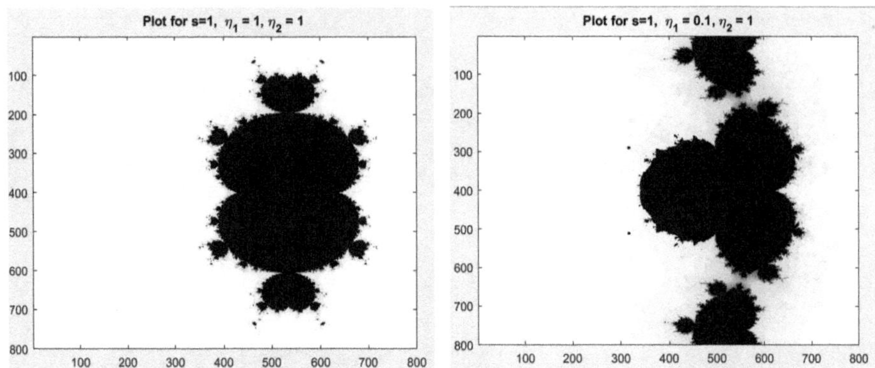

(A) Cubic Mandelbrot set generated via Picard–S iteration with s –convexity.

(B) Cubic Mandelbrot set generated via Picard–S iteration with s –convexity.

FIGURE 8.13 Cubic Mandelbrot set generated via Picard–S iteration with s –convexity for $\eta_1 = 1, \eta_2 = 1$, and $\eta_1 = 0.1, \eta_2 = 1$.

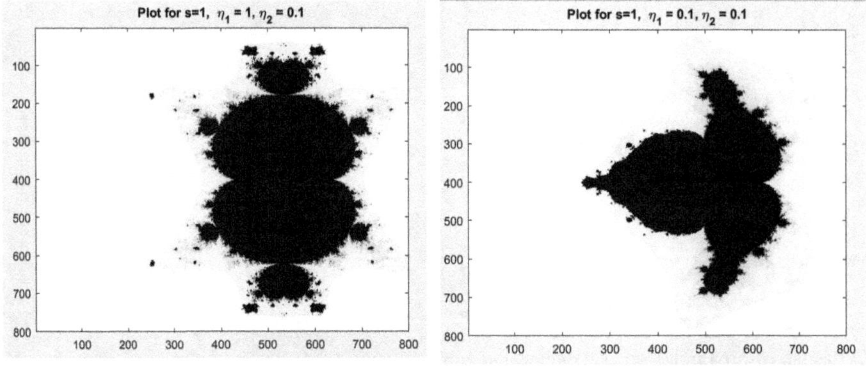

(A) Cubic Mandelbrot set generated via Picard–S iteration with s –convexity.

(B) Cubic Mandelbrot set generated via Picard–S iteration with s –convexity.

FIGURE 8.14 Cubic Mandelbrot set generated via Picard–S iteration with s –convexity for $\eta_1 = 1, \eta_2 = 0.1$, and $\eta_1 = 0.1, \eta_2 = 0.1$.

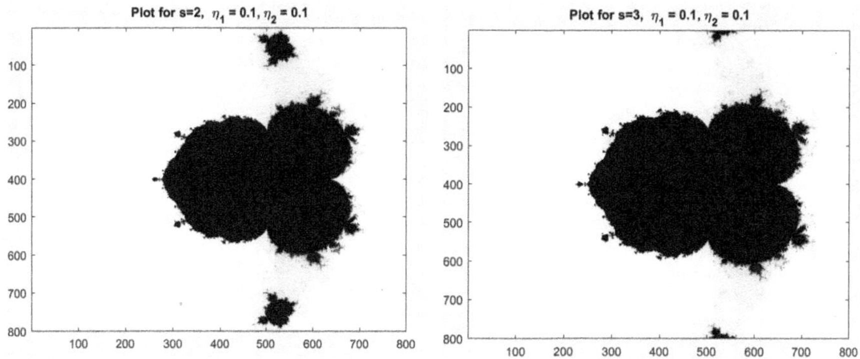

(A) Cubic Mandelbrot set generated via Picard–S iteration with s –convexity.

(B) Cubic Mandelbrot set generated via Picard–S iteration with s –convexity.

FIGURE 8.15 Cubic Mandelbrot set generated via Picard–S iteration with s –convexity for $s = 2, \eta_1 = 0.1, \eta_2 = 0.1$, and $s = 3, \eta_1 = 0.1, \eta_2 = 0.1$.

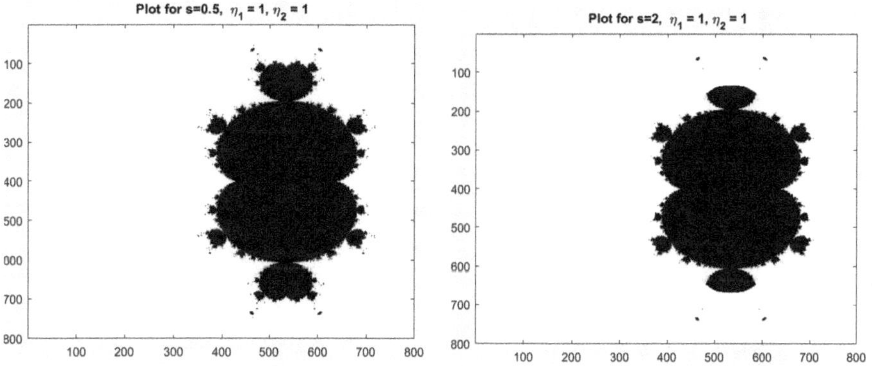

(A) Cubic Mandelbrot set generated via Picard–S iteration with s –convexity.

(B) Cubic Mandelbrot set generated via Picard–S iteration with s –convexity.

FIGURE 8.16 Quadratic Mandelbrot set generated via Picard–S iteration with s –convexity for $s = 0.5, \eta_1 = 1, \eta_2 = 1$, and $s = 2, \eta_1 = 1, \eta_2 = 1$.

For the same values of η_i, where $i = 1, 2$, but different s values for Figures 8.15(A), 8.15(B), 8.16(A), and 8.16(B) and Figures 8.17(A) and 8.17(B), we can easily analyze the fractal pattern differences.

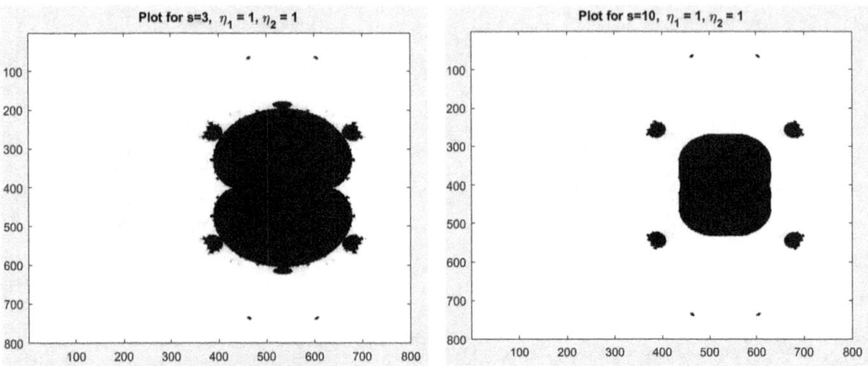

(A) Cubic Mandelbrot set generated via Picard–S iteration with s –convexity.

(B) Cubic Mandelbrot set generated via Picard–S iteration with s –convexity.

FIGURE 8.17 Cubic Mandelbrot set generated via Picard –S iteration with s – convexity for $s = 3, \eta_1 = 1, \eta_2 = 1$, and $s = 10, \eta_1 = 1, \eta_2 = 1$.

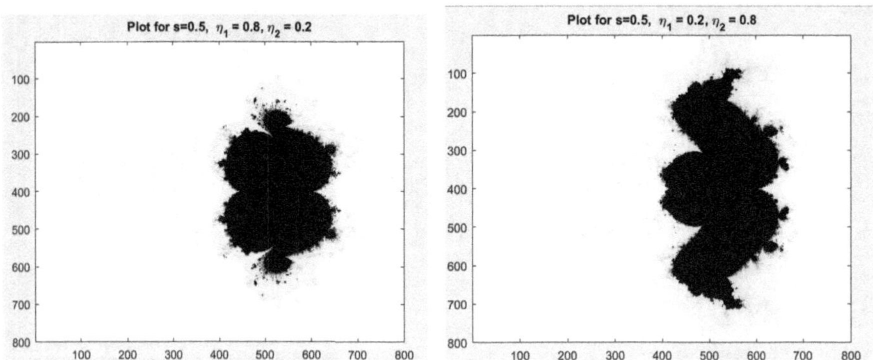

(A) Cubic Mandelbrot set generated via Picard–S iteration with s –convexity.

(B) Cubic Mandelbrot set generated via Picard–S iteration with s –convexity.

FIGURE 8.18 Cubic Mandelbrot set generated via Picard–S iteration with s –convexity for $s = 0.5, \eta_1 = 0.8, \eta_2 = 0.2$, and $s = 0.5, \eta_1 = 0.2, \eta_2 = 0.8$.

However, for different values of η_i where $i = 1,2$ and the same s values for Figures 8.18(A) and 8.18(B) and Figures 8.17(A) and 8.17(B), we can easily analyze the fractals.

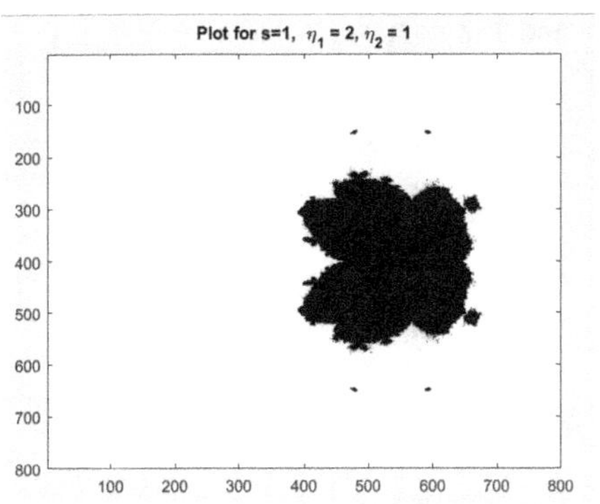

(A) Cubic Mandelbrot set generated via Picard–S iteration with s –convexity.

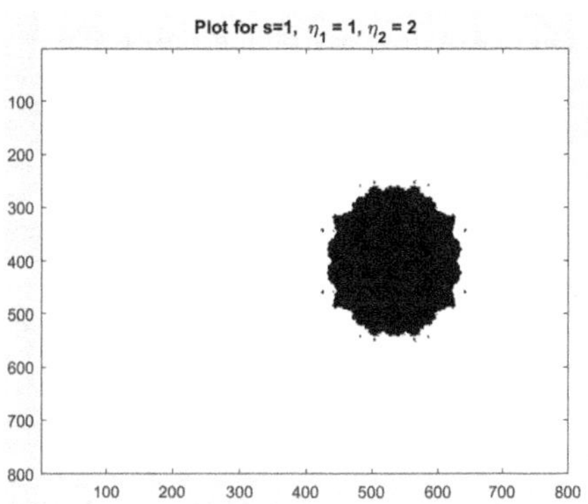

(B) Cubic Mandelbrot set generated via Picard–S iteration with s –convexity.

FIGURE 8.19 Cubic Mandelbrot set generated via Picard –S iteration with s –convexity for $s=1,\eta_1=2,\eta_2=1$, and $s=1,\eta_1=1,\eta_2=2$.

8.4 ESCAPE CRITERIA FOR HIGHER-DEGREE POLYNOMIALS

In this section, we shall obtain escape criteria for higher-degree polynomials of the form $\mathcal{Q}_\xi(\sigma)=\sigma^{\mathrm{m}+1}+\xi$, where $\mathrm{m}=1,2,3,\ldots$ in Picard–S orbit with s –convexity.

Theorem 8.2. Let $|\sigma|\geq|\xi|>\left(\frac{2}{s\eta_1}\right)^{\frac{1}{\mathrm{m}}}$ and $|\sigma|\geq|\xi|>\left(\frac{2}{s\eta_2}\right)^{\frac{1}{\mathrm{m}}}$, where $\xi\in\mathcal{C}$. Let $\ell_0=\ell$, $\sigma_0=\sigma$, and $\hbar_0=\hbar$; then sequence $\{\sigma\}$, then the sequence $\{\sigma_n\}$ is defined as

$$\begin{cases}\hbar_n &=(1-\eta_2)^s\sigma_n+\eta_2^s\mathcal{Q}_\xi(\sigma_n);\\ \ell_n &=(1-\eta_1)^s\mathcal{Q}_\xi(\sigma_n)+\eta_1^s\mathcal{Q}_\xi(\hbar_n);\\ \sigma_{n+1} &=\mathcal{Q}_\xi(\ell_n);n\geq 0\end{cases}$$

where $\eta_1,\eta_2,s\in(0,1]$, and $\mathcal{Q}_\xi(\sigma)=\sigma^{\mathrm{m}+1}+\xi$, $\mathrm{m}=1,2,3,\ldots$ Then $|\hbar_n|\to\infty$ as $n\to\infty$.

Proof. Let $|\hbar|=\left|(1-\eta_2)^s\sigma+\eta_2^s\mathcal{Q}_\xi(\sigma)\right|$. Then for $\mathcal{Q}_\xi(\sigma)=\sigma^{\mathrm{m}+1}+\xi$, $\mathrm{m}=1,2,3,\ldots$. We have

$$\begin{aligned}|\hbar| &=\left|(1-\eta_2)^s\sigma+\eta_2^s\left(\sigma^{\mathrm{m}+1}+\xi\right)\right|\\ &=\left|(1-\eta_2)^s\sigma+\left(1-(1-\eta_2)\right)^s\left(\sigma^{\mathrm{m}+1}+\xi\right)\right|\end{aligned}$$

On expanding up to linear terms of $(1-\eta_2)$ and η_2, we have

$$
\begin{aligned}
|\hbar| &= \left|(1-s\eta_2)\sigma + (1-s(1-\eta_2))\left(\sigma^{m+1}+\xi\right)\right| \\
&\geq \left|(1-s\eta_2)\sigma + s\eta_2\left(\sigma^{m+1}+\xi\right)\right| \\
&\geq \left|s\eta_2\sigma^{m+1} + (1-s\eta_2)\sigma\right| - \left|s\eta_2\xi\right| \\
&\geq \left|s\eta_2\sigma^{m+1} + (1-s\eta_2)\sigma\right| - \left|s\eta_2\sigma\right| \\
&\geq \left|s\eta_2\sigma^{m+1}\right| + \left|(1-s\eta_2)\sigma\right| - \left|s\eta_2\sigma\right| \\
&\geq \left|s\eta_2\sigma^{m+1}\right| - |\sigma| + \left|s\eta_2\sigma\right| - \left|s\eta_2\sigma\right| \\
&\geq |\sigma|\left(s\eta_2\sigma^{m} - 1\right). \qquad (8.4)
\end{aligned}
$$

It is also given that

$$|\ell| = \left|(1-\eta_1)^s \mathcal{Q}_\xi(\sigma) + \eta_1^s \mathcal{Q}_\xi(\hbar)\right|$$

which is also equivalent to

$$|\ell| = \left|(1-\eta_1)^s\left(\sigma^{m+1}+\xi\right) + (1-(1-\eta_1))^s\left(\hbar^{m+1}+\xi\right)\right|$$

since $1-s+s\eta_1 > s\eta_1$. Using binomial expansion up to linear terms of η_1 and $(1-\eta_1)$, we have

$$
\begin{aligned}
|\ell| &\geq \left|(1-s\eta_1)\left(\sigma^{m+1}+\xi\right) + (1-s(1-\eta_1))\left(\hbar^{m+1}+\xi\right)\right| \\
&\geq \left|(1-s\eta_1)\left(\sigma^{m+1}+\xi\right) + (1-s+s\eta_1)\left(\hbar^{m+1}+\xi\right)\right| \\
&\geq \left|(1-s\eta_1)\left(\sigma^{m+1}+\xi\right) + s\eta_1\left(\left(|\sigma|\left(s\eta_2|\sigma|-1\right)\right)^{m+1}+\xi\right)\right|
\end{aligned}
$$

It is given that $|\sigma| > \left(\dfrac{2}{s\eta_2}\right)^{\frac{1}{m}}$, which implies $|\sigma|^{m+1}\left(s\eta_2|\sigma|^{m}-1\right)^{m+1} > |\sigma|^{m+1}$ and $|\sigma| > \xi$, so we have

$$
\begin{aligned}
|\ell| &\geq \left|(1-s\eta_1)\left(\sigma^{m+1}+\xi\right) + s\eta_1\left(|\sigma|^{m+1}+\xi\right)\right| \\
&\geq \left|(1-s\eta_1)\sigma^{m+1} + (1-s\eta_1)\xi + s\eta_1|\sigma|_{m+1} + s\eta_1\xi\right| \\
&\geq \left|\sigma^{m+1}+\xi\right| \\
&\geq \left|\sigma^{m+1}\right| - |\xi|
\end{aligned}
$$

$$\geq \left|\sigma^{m+1}\right| - |\sigma|$$
$$\geq |\sigma|\left(|\sigma|^{m} - 1\right) \tag{8.5}$$

Similarly, $\sigma_n = \mathcal{Q}_\xi(\ell)$, which implies that $|\sigma_n| = \ell^{m+1+\xi}$. Also

$$|\sigma_1| = \left|\mathcal{Q}_\xi(\ell_n)\right|$$
$$|\sigma_1| = \left|\ell^{m+1} + \xi\right|$$
$$\geq \left|\left(|\sigma|\left(|\sigma|^2 - 1\right)\right)^{m+1}\right| - |\sigma|$$
$$\geq |\sigma|^{m+1} - |\sigma|$$
$$\geq |\sigma|\left(|\sigma|^{m} - 1\right)$$

since $|\sigma| \geq |\xi| > \left(\frac{2}{s\eta_1}\right)^{\frac{1}{m}}$. It implies that $|\sigma|^{m} - 1 > 1$. Therefore, $\exists\ \kappa > 0 \ni |\sigma|^{m} - 1 > 1 + \kappa > 1$. It results in

$$|\sigma_1| > (1-\kappa)|\sigma|$$
$$> (1-\kappa)^2 |\sigma|$$
$$> (1-\kappa)^3 |\sigma|$$

Inductively, we have

$$|\sigma_1| > (1-\kappa)^n |\sigma|$$

Hence $|\sigma|$ diverges to $+\infty$ as $n \to \infty$.

8.5 ESCAPE CRITERIA FOR COMPLEX QUADRATIC POLYNOMIALS

Theorem 8.3. Suppose that $|\sigma| \geq |\xi| > \frac{2}{s\eta_1}$ and $|\sigma| \geq |\xi| > \frac{2}{s\eta_2}$ where ξ is a complex number. Let $\ell_0 = \ell$, $\sigma_0 = \sigma$ and $\hbar_0 = \hbar$; then the sequence $\{\sigma\}$ is defined as

$$\begin{cases} \hbar_n &= (1-\eta_2)^s \sigma_n + \eta_2^s \mathcal{Q}_\xi(\sigma_n); \\ \ell_n &= (1-\eta_1)^s \mathcal{Q}_\xi(\sigma_n) + \eta_1^s \mathcal{Q}_\xi(\hbar_n); \\ \sigma_{n+1} &= \mathcal{Q}_\xi(\ell_n); n \geq 0 \end{cases}$$

where $\eta_1, \eta_2 \in (0,1]$ and $Q_\xi(\sigma) = \sigma^2 + \xi$. Then $|\sigma_n| \to \infty$ as $n \to \infty$.

Proof. Consider

$$|\hbar| = \left|(1-\eta_2)^s \sigma_n + \eta_2^s Q_\xi(\sigma_n)\right|$$

since $Q_\xi(\sigma) = \sigma^2 + \xi$. Therefore,

$$\begin{aligned}|\hbar| &= \left|(1-\eta_2)^s \sigma_n + \eta_2^s(\sigma_n^2 + \xi)\right| \\ |\hbar| &= \left|(1-\eta_2)^s \sigma_n + (1-(1-\eta_2))^s(\sigma_n^2 + \xi)\right|\end{aligned}$$

As $1 - s + s\eta_2 \geq s\eta_2$ and $|\hbar| \geq |\xi|$, so by binomial expansion up to the linear terms of η_2 and $(1-\eta_2)$, we obtain

$$\begin{aligned}|\hbar| &\geq \left|(1-s\eta_2)\sigma + (1-s(1-\eta_2))(\sigma^2+\xi)\right| \\ &\geq |(1-s\eta_2)\sigma + (1-s+s\eta_2))(\sigma^2+\xi)| \\ &\geq \left|(1-s\eta_2)\sigma + s\eta_2(\sigma^2+\xi)\right| \\ &\geq \left|(1-s\eta_2)\sigma + s\eta_2(\sigma^2+\xi)\right| \\ &\geq \left|s\eta_2\sigma^2 + (1-s\eta_2)\sigma\right| - |s\eta_2\sigma| \\ &\geq \left|s\eta_2\sigma^2\right| - \left|(1-s\eta_2)\sigma\right| - |s\eta_2\sigma| \\ &\geq \left|s\eta_2\sigma^2\right| - |\sigma| + |s\eta_2\sigma| - |s\eta_2\sigma| \\ &\geq |\sigma|\left(s\eta_2\left|\sigma^2\right| - 1\right) \end{aligned} \tag{8.6}$$

Also,

$$\begin{aligned}|\ell| &\geq \left|(1-\eta_1)^s Q_\xi(\sigma) + \eta_1^s Q_\xi(\hbar)\right| \\ &\geq \left|(1-\eta_1)^s(\sigma^2+1) + \left(1-(1-\eta_1^s)\right)(\hbar^2+1)\right|\end{aligned}$$

As $1 - s + s\eta_1 \geq s\eta_1$ and $|\hbar| \geq |\xi|$, so by binomial expansion up to the linear terms of η_1 and $(1-\eta_1)$, we obtain

$$\begin{aligned}|\ell| &\geq \left|(1-s\eta_1)(\sigma^2+\xi) + (1-s(1-\eta_1))(\hbar^2+\xi)\right| \\ &\geq \left|(1-s\eta_1)(\sigma^2+\xi) + (1-s+s\eta_1)(\hbar^2+\xi)\right| \\ &\geq \left|(1-s\eta_1)(\sigma^2+\xi) + s\eta_1\left(\left(|\sigma|(s\eta_2|\sigma|-1)\right)^2 + \xi\right)\right|\end{aligned}$$

since $|\sigma| > \dfrac{2}{\eta_1}$ implies $s\eta_1|\sigma|-1>1$ and $|\sigma|^2\left(s\eta_1|\sigma|-1\right)^2$. The previous equation can be re-written as

$$\begin{aligned}
|\ell| &\geq |(1-s\eta_1)\left(\sigma^2+\xi\right)+s\eta_1\left(|\sigma|^2+\xi\right)\\
&\geq \left|(1-s\eta_1)\sigma^2+(1-s\eta_1)\xi+s\eta_1|\sigma|^2+s\eta_1\xi\right|\\
&\geq \left|\sigma^2+\xi\right|\\
&\geq |\sigma|^2-|\xi|\\
&\geq |\sigma|^2-|\sigma|\\
&\geq |\sigma|\left(|\sigma|-1\right).
\end{aligned}$$

Considering the fact that $|\sigma_1| = \mathcal{Q}_\xi(\ell) \geq \left|\ell^2+\xi\right|$, we have

$$\begin{aligned}
|\sigma_1| &\geq \left||\ell|^2-\xi\right|\\
&\geq \left|(\sigma_2+\xi)^2-\xi\right|\\
&\geq \left|\left(|\sigma|\left(|\sigma|-1\right)\right)^2-\xi\right|\\
&\geq \left|\left(|\sigma|\left(|\sigma|-1\right)\right)^2-\sigma\right|\\
&\geq \left|\left(|\sigma|\left(|\sigma|-1\right)\right)^2\right|-|\sigma|\\
&\geq |\sigma|\left(|\sigma|\left(\left(|\sigma|-1\right)\right)^2-1\right)\\
&\geq |\sigma|\left(|\sigma|-1\right).
\end{aligned}$$

Also, $\hbar_1 = (1-\eta_2)^s\sigma+\eta_2^s\mathcal{Q}_\xi(\sigma)$ implies $|\hbar_1| = \left|(1-\eta_2)^s\sigma+\left(1-(1-\eta_2)\right)^s\left(\sigma^2+\xi\right)\right|$. Since $(1-s+s\eta_2)$ and $|\sigma|\geq|\xi|$, using the binomial expansion up to the linear terms of η_2^s and $\left(1-\eta_2^s\right)$, we have

$$\begin{aligned}
|\hbar_1| &= \left|(1-s\eta_2)\sigma+\left(1-s(1-\eta_2)\right)\left(\sigma^2+\xi\right)\right|\\
&\geq \left|(1-s\eta_2)\left(|\sigma|\left(|\sigma|-1\right)\right)+(1-s+s\eta_2)\left(\left(|\sigma|\left(|\sigma|-1\right)\right)^2+\xi\right)\right|\\
&\geq \left|(1-s\eta_2)|\sigma|+(1-s+s\eta_2)\left(|\sigma|^2+\xi\right)\right|\\
&\geq \left|(1-s\eta_2)|\sigma|+s\eta_2\left(|\sigma|^2+\xi\right)\right|
\end{aligned}$$

$$\geq \left|(1-s\eta_2)|\sigma| + s\eta_2|\sigma|^2\right| - |s\eta_2\xi|$$
$$\geq \left|s\eta_2|\sigma|^2 + (1-s\eta_2)|\sigma|\right| - |s\eta_2\sigma|$$
$$\geq \left|s\eta_2|\sigma|^2\right| - \left|(1-s\eta_2)|\sigma|\right| - |s\eta_2\sigma|$$
$$\geq s\eta_2|\sigma|_2| - |\sigma| + s\eta_2|\sigma| - |s\eta_2\sigma|$$
$$\geq |\sigma|(s\eta_2|\sigma| - 1)$$

It is given that $|\sigma| > \frac{2}{s\eta_2}$, which implies $s\eta_2|\sigma| > 1$; there is a number $\kappa > 0$ such that $s\eta_2|\sigma| - 1 > 1 + \kappa > 1$. Therefore

$$|\sigma_1| > (1+\kappa)|\sigma|$$
$$|\sigma_2| > (1+\kappa)^2|\sigma|$$
$$|\sigma_3| > (1+\kappa)^3|\sigma|$$
$$|\sigma_4| > (1+\kappa)^4|\sigma|$$

Inductively, we have

$$|\sigma_n| > (1+\kappa)^n|\sigma|$$

It results, in $|\sigma_n| \to \infty$ as $n \to \infty$.

Corollary 8.1. Let $|\sigma| \geq |\xi| > \frac{2}{s\eta_1}$ and $|\sigma| \geq |\xi| > \frac{2}{s\eta_2}$; then the orbit $P-SSO(Q_\xi, 0, s\eta_1, s\eta_2)$ escapes to infinity.

Corollary 8.2. (Escape criterion) Let $|\sigma| > max\left\{|\xi|, \frac{2}{s\eta_1}, \frac{2}{s\eta_2}\right\}$, then $|\sigma_n| \geq (1+\kappa)^n|\sigma|$, and $|\sigma_\eta| \to \infty$ as $\eta \to \infty$.

8.6 ESCAPE CRITERIA FOR CUBIC POLYNOMIALS

Now, we study the results for the cubic polynomial $Q_\xi(\sigma) = \sigma^3 + c$, where $c \in C$ (set of all complex numbers), in Picard $-S$ orbit with $s-$convexity.

Theorem 8.4. Let $|\sigma| \geq |\xi| > \left(\frac{2}{s\eta_1}\right)^{\frac{1}{2}}$ and $|\sigma| \geq |\xi| > \left(\frac{2}{s\eta_2}\right)^{\frac{1}{2}}$, where $\xi \in C$. Let $\ell_0 = \ell$, $\sigma_0 = \sigma$, and $\hbar_0 = \hbar$; then the sequence $\{\sigma\}$ is defined as

$$\begin{cases} \hbar_n &= (1-\eta_2)^s\sigma_n + \eta_2^s Q_\xi(\sigma_n); \\ \ell_n &= (1-\eta_1)^s Q_\xi(\sigma_n) + \eta_1^s Q_\xi(\hbar_n); \\ \sigma_{n+1} &= Q_\xi(\ell_n); n \geq 0 \end{cases}$$

where $\eta_1, \eta_2, s \in (0,1]$, and $\mathcal{Q}_\xi(\sigma) = \sigma^3 + \xi$. Then $|\sigma_n| \to \infty$ as $n \to \infty$.

Proof. Let $|\hbar| = \left|(1-\eta_2)^s \sigma_n + \eta_2^s \mathcal{Q}_\xi(\sigma_n)\right|$. Now, for $\mathcal{Q}_\xi(\sigma_n)$ we have

$$
\begin{aligned}
\mathcal{Q}_\xi(\sigma_n) &= (\sigma_n)^3 + \xi \\
|\hbar| &= \left|(1-\eta_2)^s \sigma_n + \eta_2^s \sigma_n^3 + \xi\right| \\
|\hbar| &= \left|(1-\eta_2)^s \sigma_n + \left(1-(1-\eta_2)\right)^s \sigma_n^3 + \xi\right|
\end{aligned}
$$

As $1-s+s\eta_2 \geq s\eta_2$ and $|\sigma| \geq |\xi|$, by binomial expansion up to THE linear terms of η_2 and $(1-\eta_2)$, we obtain

$$
\begin{aligned}
|\hbar| &\geq \left|(1-s\eta_2)\sigma + \left(1-s(1-\eta_2)\right)\left(\sigma^3+\xi\right)\right| \\
&\geq |(1-s\eta_2)\sigma + (1-s+s\eta_2))\left(\sigma^3+\xi\right)| \\
&\geq \left|(1-s\eta_2)\sigma + s\eta_2\left(\sigma^3+\xi\right)\right| \\
&\geq \left|(1-s\eta_2)\sigma + s\eta_2\left(\sigma^3+\xi\right)\right| \\
&\geq |s\eta_2)\sigma^3 + (1-s\eta_2)\sigma| - |s\eta_2\sigma| \\
&\geq |s\eta_2)\sigma^3| - |(1-s\eta_2)\sigma| - |s\eta_2\sigma| \\
&\geq |s\eta_2)\sigma^3| - |\sigma| + |s\eta_2\sigma| - |s\eta_2\sigma| \\
&\geq |\sigma|\left(s\eta_2|\sigma|^2 \quad 1\right)
\end{aligned}
$$

and

$$
\begin{aligned}
|\ell| &\geq \left|(1-\eta_1)^s \mathcal{Q}_\xi(\sigma) + \eta_1^s \mathcal{Q}_\xi(\hbar)\right| \\
&\geq \left|(1-\eta_1)^s\left(\sigma^3+\xi\right) + \left(1-(1-\eta_1)\right)^s\left(\hbar^3+\xi\right)\right|
\end{aligned}
$$

As $1-s+s\eta_1 \geq s\eta_1$ and $|\hbar| \geq |\xi|$, by binomial expansion up to the linear terms of η_1 and $(1-\eta_1)$, we obtain

$$
\begin{aligned}
|\ell| &\geq \left|(1-s\eta_1)\left(\sigma^3+\xi\right) + \left(1-s(1-\eta_1)\right)\left(\hbar^3+\xi\right)\right| \\
&\geq |(1-s\eta_1)\left(\sigma^3+\xi\right) + (1-s+s\eta_1))\left(\hbar^3+\xi\right)| \\
&\geq \left|(1-s\eta_1)\left(\sigma^3+\xi\right) + s\eta_1\left|\left(|\sigma|\left(s\eta_2|\sigma|^2-1\right)\right)^3+\xi\right|\right|
\end{aligned}
$$

Since $|\sigma| > \left(\frac{2}{\eta_1}\right)^{\frac{1}{2}}$ implies $(s\eta_1|\sigma|^2 - 1 > 1)$ and $|\sigma|^3\left(s\eta_1|\sigma|^2 - 1\right)^3$. Also, $\sigma \geq |\xi|$.
Hence, above equation can be re – written as –

$$
\begin{aligned}
|\ell| &\geq \left|(1 - s\eta_1)(\sigma^3 + \xi) + s\eta_1\left((|\sigma|^3 + \xi\right)\right| \\
&\geq \left|(1 - s\eta_1)\sigma^3 + (1 - s\eta_1)\xi + s\eta_1|\sigma|^3 + s\eta_1\xi\right| \\
&\geq |\sigma^3 + \xi| \\
&\geq |\sigma|^3 - |\xi| \\
&\geq |\sigma|^3 - |\sigma| \\
&\geq |\sigma|\left(|\sigma|^2 - 1\right)
\end{aligned}
$$

Also,

$$
\begin{aligned}
|\sigma_1| = \mathcal{Q}_\xi(\ell) &\geq |\ell^3 + \xi| \\
&\geq |\ell^3 + \xi| \\
&\geq \left|(\sigma_2 + \xi)^3 - \xi\right| \\
&\geq \left|(|\sigma|(|\sigma| - 1))^3 - \xi\right| \\
&\geq \left|(|\sigma|(|\sigma| - 1))^3 - \sigma\right| \\
&\geq \left|(|\sigma|(|\sigma| - 1))^3\right| - |\sigma| \\
&\geq |\sigma|\left(|\sigma|(|\sigma| - 1|)^3 - 1\right) \\
&\geq |\sigma|\left(|\sigma|(|\sigma| - 1|)^3 - 1\right) \\
&\geq |\sigma|\left(|\sigma|^2 - 1\right)
\end{aligned}
$$

It is given that, $|\sigma| > \frac{2}{s\eta_2}$ which implies $s\eta_2|\sigma| > 1$, there is an existence of a number $\kappa > 0$, in such a way that $s\eta_2|\sigma|^2 - 1 > 1 + \kappa > 1$. Therefore

$$\begin{aligned}|\sigma_1| &> (1+\kappa)|\sigma| \\ |\sigma_2| &> (1+\kappa)^2|\sigma| \\ |\sigma_3| &> (1+\kappa)^3|\sigma| \\ |\sigma_4| &> (1+\kappa)^4|\sigma|\end{aligned}$$

Inductively, we have

$$|\sigma_n| > (1+\kappa)^n|\sigma|$$

It results, $|\sigma_n| \to \infty$ as $n \to \infty$.

Corollary 8.3. (Escape criterion) Let $|\sigma| > max\left\{|C|, \left(\frac{2}{s\eta_1}\right)^{\frac{1}{2}}, \left(\frac{2}{s\eta_2}\right)^{\frac{1}{2}}\right\}$; then $|\sigma_n| \to \infty$ as $n \to \infty$ when $Q_\xi(\sigma) = \sigma^3 + \xi$ and ξ is a complex constant.

Conclusion. Nature's approximate fractals show self-similarity over wide but constrained scale ranges. The relationship between fractals and leaves, in particular, is currently being utilized to estimate the amount of carbon stored in trees. Fractal analysis was reportedly utilized to distinguish real from counterfeit Pollocks with an approximately 95% success rate. Pollock's fractals reduce stress in observers in much the same manner that digitally created fractals and nature's fractals do, according to cognitive neuroscientists. Fractal antennas, fractal transistors, fractal heat exchangers, digital imaging architecture, and urban expansion are all examples of how fractals are applied in technology. In this chapter, we described how to use the Picard$-S$ iteration method with $s-$convexity to generate Mandelbrot sets using quadratic, cubic, and $(k+1)^{th}$ degree polynomials. We studied distinct fractals for complex functions in the Picard$-S$ orbit with $s-$convexity that are completely different from those existing in the literature. The results obtained are $s-$convexity applications. We generated fascinating Mandelbrots by varying the values of η_1, η_2, η_3, and s. Mandelbrot sets for complex quadratic, cubic, and $(k+1)^{th}$ degree polynomials have been shown in a few cases. Mandelbrot sets for varying values of η_1, η_2, and η_3 show quite significant changes.

REFERENCES

[1] B. Mandelbrot, *The Fractal Geometry of Nature*, W. H. Freeman and Co., San Francisco, 1982.

[2] D. Ashlock, Evolutionary exploration of the Mandelbrot set, *IEEE Cong. Evol. Comput.*, vol. 2006, pp. 2079–2086, 2006.

[3] C. A. Pickover, *Computers, Pattern, Chaos, and Beauty*, St. Martin's Press, New York, 1990.

[4] Y. C. Kwun, A. A. Shahid, W. Nazeer, M. Abbas and S. M. Kang, Fractal generation via CR iteration scheme with S-convexity, *IEEE Access*, vol. 7, pp. 69986–69997, 2019.

[5] B. Branner and J. Hubbard, The iteration of cubic polynomials I, *Acta. Math.*, vol. 160, no. 3–4, pp. 143–206, 1988.

[6] A. A. Shahid, W. Nazeer and K. Gdawiec, The Picard–Mann iteration with s-convexity in the generation of Mandelbrot and Julia sets, *Monatsh. Math.*, vol. 195, pp. 565–584, 2021.

[7] A. J. Crilly, R. A. Earnshaw and H. Jones, *Fractals and Chaos*, Springer-Verlag, New York, 1991.

[8] W. D. Crowe, R. Hasson, P. J. Rippon and P. E. Strain-Clark, On the structure of the Mandelbar set, *Nonlinearity*, vol. 2, no. 4, pp. 541–553, 1989.

[9] R. M. Crownover, *Introduction to Fractals and Chaos*, Jones Barlett Publ., 1995.

[10] R. L. Devaney, *An Introduction to Chaotic Dynamical Systems*, The Benjamin/ Cummings Publ. Co. Inc., CA, 1986.

[11] R. L. Devaney, *A First Course in Chaotic Dynamical Systems: Theory and Experiment*, Addison-Wesley Publ. Co, 1992.

[12] J. Kigami, *Analysis on Fractals*, Cambridge University Press, Cambridge, 2001.

[13] H. O. Peitgen, H. Jürgens and D. Saupe, *Chaos and Fractals*, Springer-Verlag, New York, 1992.

[14] H. O. Peitgen and D. Saupe (eds), *The Science of Fractal Images*, Springer-Verlag, New York, 1988.

[15] J. Argyris, I. Andreadis and T. E. Karakasidis, On perturbation of the Mandelbrot map, *Chaos Solitons Fract.*, vol. 11, no. 7, 2000.

[16] M. Rani and V. Kumar, Superior Julia sets, *J. Korea Soc. Math. Educ. Seri. D Res. Math. Educ.*, vol. 8, no. 4, pp. 261–277, 2004.

[17] R. Rana, Y. S. Chauhan and A. Nagi, Non linear dynamics of Ishikawa iteration, *Inter. J. Computer Appl.*, vol. 7, no. 13, pp. 43–49, 2010.

[18] G. Julia, Memoire sur l'iteration des fonctions rationnelles, *J. Math. Pures Appl.*, vol. 8, pp. 245–247, 1918.

[19] H. O. Peitgen, E. Maletsky, and H. Jargens, T. Perciante, D. Saupe and L. Yunker, Chaos, in *Fractals for the Classroom: Strategic Activities Volume Two*, Springer, New York, 1992, pp. 51–106.

[20] B. B. Mandelbrot, *The Fractal Geometry of Nature*, WH Freeman, New York, 1982, vol. 2.

[21] Y. S. Chauhan, R. Rana and A. Negi, New Julia sets of Ishikawa iterates, *Int. J. Comput. Appl.*, vol. 7, no. 13, pp. 34–42, 2010.

[22] S. M. Kang, A. Rafiq, A. Latif, A. A. Shahid and F. Ali, Fractals through modified iteration scheme, *Filomat*, vol. 30, no. 11, pp. 3033–3046, 2016.

[23] S. M. Kang, A. Rafiq, A. Latif, A. A. Shahid and Y. C. Kwun, Tricorns and multicorns of s –iteration scheme, *J. Function Spaces*, vol. 2015, Art. no. 417167, 2015.

[24] Ashish, M. Rani and R. Chugh, Julia sets and Mandelbrot sets in Noor orbit, *Appl. Math. Comput.*, vol. 228, pp. 615–631, 2014.

[25] F. Bernstein and G. Doetsch, Zur theorie der konvexen funktionen, *Math. Ann.*, vol. 76, no. 4, pp. 514–526, 1915.

[26] W. W. Breckner and G. Orbãn, *Continuity Properties Rationally s –Convex Mappings with Values Ordered Topological Linear Space*, Babe s –Bloyai Univ., Cluj-Napoca, Romania, Tech. Rep., 1978.

[27] M. Alomari and M. Darus, Co-ordinated s –convex function in the first sense with some hadamard-type inequalities, *Int. J. Contemp. Math. Sci.*, vol. 3, no. 32, pp. 1557–1567, 2008.

[28] U. S. Kirmaci, M. K. Bakula, M. E. Ozdemir, and J. Pecaric, Hadamard-type inequalities for s –convex functions, *Appl. Math. Comput.*, vol. 193, no. 1, pp. 26–35, 2007.

[29] H. Hudzik and L. Maligranda, Some remarks on s –convex functions, *Aequ. Math.*, vol. 48, no. 1, pp. 100–111, 1994.

[30] W. Takahashi, A convexity in metric space and nonexpansive mappings, *Dept. Math., Tokyo Inst. Technol., Tokyo, Japan, Kodai Math. Seminar Rep.*, vol. 22, pp. 142–149, 1970. https://doi.org/10.2996/kmj/1138846111

[31] M. R. Pinheiro, s –convexity (foundations for analysis), *Differ. Geom. Dyn. Syst.*, vol. 10, pp. 257–262, 2008.

[32] M. K. Mishra, D. B. Ojha and D. Sharma, Some common fixed point results in relative superior Julia sets with Ishikawa iteration and sconvexity, *Int. J. Adv. Eng. Sci. Technol.*, vol. 2, no. 2, pp. 175–180, 2011.

[33] M. K. Mishra, D. B. Ojha and D. Sharma, Fixed point results in tricorn and multicorns of Ishikawa iteration and s –convexity, *Proc. Int. J. Adv. Eng. Sci. Technol.*, vol. 2, no. 2, pp. 157–160, 2011.

[34] S. M. Kang, W. Nazeer, M. Tanveer, and A. A. Shahid, New fixed point results for fractal generation in Jungck Noor orbit with s –convexity, *J. Function Spaces*, vol. 2015, Art. no. 963016, 2015.

[35] Y. C. Kwun, M. Tanveer, W. Nazeer, M. Abbas and S. M. Kang, Fractal generation in modified jungck-S orbit, *IEEE Access*, vol. 7, pp. 35060–35071, 2019.

[36] R. Chugh, V. Kumar and S. Kumar, Strong convergence of a new three step iterative scheme in Banach spaces, *Amer. J. Comput. Math.*, vol. 2, no. 4, p. 345, 2012.

[37] R. Agarwal, D. O Regan and D. Sahu, Iterative construction of fixed points of nearly asymptotically nonexpansive mappings, *J. Nonlinear Convex Anal.*, vol. 8, pp. 61–79, 2007.

[38] E. Picard, Memoire sur la theorie des equations aux derivees partielles et la methode des approximations successives, *J. Math. Pures Appl.*, vol. 6, pp. 145–210, 1890.

[39] R. Agarwal, D. O Regan and D. Sahu, Iterative construction of fixed points of nearly asymptotically nonexpansive mappings, *J. Nonlinear Convex Anal.*, vol. 8, pp. 61–79, 2007.

[40] D. R. Sahu, Applications of the S-iteration process to constrained minimization problems and split feasibility problems, *Fixed Point Theory*, vol. 12, pp. 187–204, 2011.

[41] W. R. Mann, Mean value methods in iteration, *Proc. Am. Math. Soc.*, vol. 4, pp. 506–510, 1953.

[42] R. Chugh and V. Kumar, Stability of hybrid fixed point iterative algorithms of Kirk-Noor type in normed linear space for self and nonself operators, *Int. J. Contemp. Math. Sci.*, vol. 7, pp. 1165–1184, 2012

[43] S. Ishikawa, Fixed points by a new iteration method, *Proc. Am. Math. Soc.*, vol. 44, pp. 147–150, 1974.

[44] M. A. Noor, New approximation schemes for general variational inequalities. *J. Math. Anal. Appl.*, vol. 251, pp. 217–229, 2000.

[45] W. Phuengrattana and S. Suantai, On the rate of convergence of Mann, Ishikawa, Noor and SP-iterations for continuous functions on an arbitrary interval, *J. Comput. Appl. Math.*, vol. 235, pp. 3006–3014, 2011.

[46] F. Gursoy and V. Karakaya, A $picard - s$ hybrid type iteration method for solving a differential equation with retarded argument, *arXiv:1403.2546v2* [math.FA], 28 Apr 2014.

[47] M. F. Barnsley, *Fractals Everywhere*, Academic, New York, 2014.

[48] R. L. Devaney, *A First Course in Chaotic Dynamical Systems: Theory and Experiment*, Addison-Wesley, New York, 1992.

[49] L. Xiang-dong, Y. De-jun, Z. Wei-yong and W. Guang-xing, The bounds of the general M and J sets and the estimations for the Hausdorff's dimension of the general, *J. Set. Appl. Math. Mech.*, vol. 22, no. 11, pp. 1318–1324, 2001.

[50] S. H. Khan, A Picard–Mann hybrid iterative process, *Fixed Point Theory Appl.*, vol. 2013, Article ID 69, pp. 1–10, 2013.

Index

For Product Safety Concerns and Information please contact our EU representative GPSR@taylorandfrancis.com Taylor & Francis Verlag GmbH, Kaufingerstraße 24, 80331 München, Germany

Batch number: 10397790

Printed by Printforce, the Netherlands